Developing Baseline Communication Skills

Developing Baseline Communication Skills

Catherine Delamain & Jill Spring

Speechmark Publishing
Telford Road • Bicester • Oxon OX26 4LQ • UK

> Please note that in this text, for reasons of clarity alone,
> 'he' is used to refer to the child and 'she' to the teacher.

Published by
Speechmark Publishing Ltd, Telford Road, Bicester, Oxon OX26 4LQ, UK

www.speechmark.net

© Catherine Delamain & Jill Spring, 2000

First published 2000
Reprinted 2002, 2003, 2004, 2005

All rights reserved. The whole of this work, including all text and illustrations is protected by copyright. No part of it may be copied, altered, adapted or otherwise exploited in any way without express prior permission, unless it is in accordance with the provisions of the Copyright Designs and Patents Act 1988 or in order to photocopy or make duplicating masters of those pages so indicated, without alteration and including copyright notices, for the express purposes of instruction and examination. No parts of this work may otherwise be loaded, stored, manipulated, reproduced, or transmitted in any form or by any means, electronic or mechanical, including photocopying and recording, or by any information, storage and retrieval system without prior written permission from the publisher, on behalf of the copyright owner.

002-4757/Printed in the United Kingdom/1030

British Library Cataloguing in Publication Data
Delamain, Catherine
　　Developing baseline communication skills
　　1. English language – Study and teaching (Preschool)
　　2. Communicative competence in children – Study and teaching (Preschool)
　　I. Title   II. Spring, Jill
　　155.4'136

ISBN 0 86388 481 4
(Previously published by Winslow Press Ltd under ISBN 0 86388 277 3)

Contents

Page
- vi / Acknowledgements
- vii / Preface
- 1 / Introduction
- 3 / How to use this Book

PERSONAL AND SOCIAL DEVELOPMENT ACTIVITIES

- 7 / Turn Taking
- 29 / Body Language
- 51 / Awareness of Others
- 73 / Confidence and Independence
- 95 / Feelings and Emotions

LANGUAGE AND LITERACY ACTIVITIES

- 117 / Understanding
- 139 / Listening and Attention
- 161 / Speaking
- 183 / Auditory Memory
- 205 / Phonological Awareness

ACTIVITY RESOURCES

- 227 / Resources
- 281 / Cross-Reference Tables
- 293 / Pupil Record Sheets

Acknowledgements

The authors would like to thank the speech and language therapy colleagues with whom they have shared the fun of working with groups of children. All such colleagues will have contributed, wittingly or unwittingly, to this collection of games, as old ideas are adapted and improved, and new ones invented.

They would also like to thank all the teachers with whom they have worked, particularly those language unit teachers with whom they have been most closely involved.

Preface

Baseline Assessments are now mandatory for all children entering reception classes in state schools. The assessments look at the child's personal and social development, his language skills, and his readiness for reading and writing. A child may be graded from Level I to Level IV in any of the identified developmental or skill areas, with Level IV representing the point at which the child is deemed ready to embark on the formal education of Key Stage I.

Reception teachers are thoroughly accustomed to making formal or informal appraisals of new arrivals in their class, and to planning individual curricula to meet individual needs. However, the Baseline Assessments give this appraisal a more formal structure, and perhaps identify some developmental areas that have not in the past been allocated specific attention in the curriculum. The authors of this book, both speech & language therapists with a long history of involvement in education, anticipate a demand for teaching resources to meet these new needs. This collection of simple games is the result.

Teachers using the Baseline Assessments will quickly identify those children whose development in one or more areas lags behind the level expected for their age. There are likely to be groups with similar needs in the same areas. Teachers can turn to the appropriate sections of this resource book, and select suitable games for those groups. We are keenly aware that teachers are reluctant to adopt resources involving further testing, complicated paperwork, expensive equipment, or special slots in an already crowded curriculum. The use of this pack therefore has the following advantages:

- No testing other than Baseline Assessments.
- Equipment not needed or kept to a minimum.
- Games can be fitted in to the existing curriculum (circle time, literacy hour, outdoor play, hall and PE, small group work in the classroom).
- Games can mostly be organised by classroom assistants or volunteers.
- Record keeping is simple and minimal.

We hope that this resource will prove a useful tool.

Introduction

Over the past few years there has been an apparent deterioration in the communication skills of young children. Teachers have expressed concern that children entering reception classes frequently lack the listening, understanding and speaking skills necessary if they are to make a happy and successful start to their school careers.

Language is the medium by which education is chiefly delivered. This is acknowledged in the National Framework for Baseline Assessment, with its emphasis on listening, responding and speaking. Language is also crucial to social relationships, and therefore forms an important part of children's personal and social development. Finally, many of the skills underlying speech and language competence are the same as those needed for learning to read, write and spell.

Children may now enter school at any age from only just four, to five. There will inevitably be extremely wide variations in their speech and language skills, with some of them, particularly the youngest four-year-olds, being unprepared for the language demands which will be put on them.

> *Language is the medium by which education is chiefly delivered. This is acknowledged in the National Framework for Baseline Assessment, with its emphasis on listening, responding and speaking.*

Baseline Assessments look at a wider range of areas than is addressed in this pack. The activities included here focus on social communication and certain pre-literacy skills, helping to move reception-class children towards the desirable learning outcomes identified in Baseline Stage IV. Levels I–IV in this book correspond to Levels I–IV in the Baseline Assessment.

The activities will also be useful for four- to five-year-old children in nursery education and in playgroups, and for groups in speech & language therapy clinics. In these cases the appropriate level at which to start a child will be decided by whatever assessments are in use, and by observation.

> **NOTE:**
> Variations on some of the games and activities included in this pack are in common use among speech & language therapists, and are also used in some educational settings. If we have inadvertently included a game which somebody feels originated with them, or have used a name which already exists, we can only apologise.

This pack consists of 200 games and activities for whole classes or groups. All the activities can be incorporated easily into the curriculum, and suggestions are given as to where they would fit in most appropriately. Every effort has been made to give clear, explicit instructions for playing the games. Equipment needed has been kept to a minimum. Where equipment is involved, it can either readily be assembled in the classroom, or may be photocopied from the Activity Resources section of this book. In a few instances, the templates in the Resources will need to be coloured after photocopying.

The activities are broadly classified under Personal and Social Development and Language and Literacy Development. The Personal and Social Development section covers turn taking, body language, awareness of others, confidence and independence. The Language and Literacy section covers understanding, listening and attention, speaking, auditory memory and phonological awareness.

Teachers will notice that there is considerable emphasis on the development of child-to-child talk. An extremely high percentage of talk in the classroom consists of adult-to-child talk. A much smaller percentage consists of child-to-adult talk. Child-to-child talk does, of course, take place while the children are working in their groups or at meals and playtimes, but has not been recognised as an area requiring a structured programme of development. For children who are naturally good communicators this is not a problem. For children with poor language skills, help in learning to communicate effectively with their peers is very important.

One other area addressed under phonological awareness may be unfamiliar to teachers. Recent research has shown that the ability to speak quickly while maintaining clarity is, amongst many others, a predictor of good progress with literacy. We have therefore included a small selection of 'speed speech' activities, and feel sure that teachers, with their usual inventiveness, will be able to think of many more.

We hope that this book will prove easy to use, useful, and fun for the children.

> *Recent research has shown that the ability to speak quickly while maintaining clarity is, amongst many others, a predictor of good progress with literacy.*

How to use this Book

LAYOUT

Developing Baseline Communication Skills is divided into three sections – two activity sections, Personal and Social Development, and Language and Literacy – and Activity Resources. At the beginning of each of the two main activity sections there is a contents page listing the five skill areas covered in that section.

Each skill area consists of 20 activities divided into four levels, which broadly correspond to Levels I–IV in the Baseline Assessment scheme. The activities are listed at the beginning of the skills areas.

Every activity sheet includes an explanation of its aim, the equipment needed, and instructions on how to play the game. The tabs on each sheet indicate the level of the game, and suggest appropriate curriculum areas in which to carry it out.

THE SKILL AREAS

Personal and Social Development

▲ *Turn-Taking*
Turn-taking involves the ability to participate in cooperative games, the ability to wait one's turn in a game, the ability to take conversational turns, and the ability to contribute appropriately in classroom discussions.

▲ *Body Language*
Body language includes facial expression, gesture, posture and proximity, and the ability to recognise these signals in others.

▲ *Awareness of Others*
Awareness of others involves the recognition of similarities and differences between people, including their individuality, needs and thoughts.

▲ *Confidence and Independence*
Confidence and independence include the child's ability to move around the school environment, and to complete simple tasks with the minimum of adult support.

▲ *Feelings and Emotions*
This skill area focuses on the concepts of basic emotions, and the language used to express them.

Language and Literacy

▲ *Understanding*
This refers to the ability to derive meaning from spoken language. It covers vocabulary, instructions, questions, explanations, stories and conversations.

▲ *Listening and Attention*
This refers to the ability to focus and pay attention during spoken language activities, and maintain sufficient concentration.

> **NOTE:**
> Poor listening and attention skills will affect understanding.

▲ *Speaking*
This refers to the ability of the child to express his needs and ideas in coherent spoken language. There is an emphasis in this pack on the development of child-to-child talk.

▲ *Auditory Memory*
This refers to the ability to retain short pieces of spoken information for long enough to process them.

> **NOTE:**
> The needs of children with significant language and/or speech deficits are not addressed in this volume.

▲ *Phonological Awareness*
Phonological awareness refers to the range of metalinguistic skills necessary for the development of literacy, including rhyme awareness, syllabification, alliteration and phonemic knowledge.

ESTABLISHING STARTING LEVELS

It is likely that children will be at different levels in different areas. For instance, a child with poor listening and attention skills, but with relatively good social communication skills, may need to do activities at Level I in the Language and Literacy section, but at Level III or IV in the Personal and Social section. A simple record sheet is included in the Activity Resources section. Information derived from the Baseline Assessment may be transferred to the record sheet to help in deciding appropriate starting levels in the various skill areas.

The levels in the 10 skill areas correspond broadly to developmental criteria as follows:

Level I	3.11 – 4.05 years
Level II	4.03 – 4.09 years
Level III	4.07 – 5.01 years
Level IV	4.11 – 5.05 years

In most cases a child's performance according to the Baseline Assessment criteria will match the corresponding levels in this book. However, it is vital to remember the following:

> Development of understanding precedes speaking; a child's understanding must therefore be **at least at the same level as or higher than the speaking level targeted**.
>
> Development of understanding and speaking must be **at least at the same level or higher than the phonological awareness level targeted**.

MOVING FORWARDS OR BACKWARDS

It is important not to move on too quickly. If the activity is too difficult the child will start to experience failure, and this must be avoided in order to maintain self-esteem and build confidence. There is a certain amount of overlap between the various sections, and by looking at the cross-reference tables (Activity Resources) it is possible to extend a certain type of activity laterally, before moving on.

If a child experiences difficulty with an activity, go back to the previous level. Continue at this level until you are confident that he is ready to move on.

PRACTICAL ISSUES

Activity Resources

Several of the activities involve the use of pictorial material, texts or word lists. These may be photocopied from the Activity Resources section, and the pictures coloured as directed. Teachers may invent their own additional material or use school resources using the samples as a guideline.

▲ *Zig*
  Zig the Alien features in several of the activities. Zig could be made as a sock or glove puppet, or a commercial soft toy may be used. It is important to maintain his identity and use the same puppet/doll each time he appears.

▲ *Pronunciation of phonemes*
  When pronouncing the following single phonemes, do not add 'uh', eg, 'ssss' not 'suh': (m,n,k,t,p,f,sh,s,z,ch,l,th,h).

Developing Baseline Communication Skills
PERSONAL AND SOCIAL DEVELOPMENT

Level I

8 / Musical Hat
9 / Talking Toy (i)
10 / Group Lotto
11 / Balloon Bubbles (i)
12 / The Farmer Wants a Horse

Level II

13 / Talking Toy (ii)
14 / Pull out a Name
15 / Number's Up!
16 / Add To It
17 / Nursery Rhyme Circle

Level III

18 / Feely Bag
19 / Yes-No
20 / In My Case
21 / Build It
22 / Balloon Bubbles (ii)

Level IV

23 / Talking Toy (iii)
24 / 30 Seconds
25 / Balloon Bubbles (iii)
26 / Story Line
27 / Post Box

PERSONAL AND SOCIAL DEVELOPMENT

Musical Hat

Aim
To be able to respond to musical cues.

Any kind of hat, the funnier the better.
Tape recorder and music tape, or piano.

Children sit in a circle. The hat is given to the first child, and he is told to start passing it round the circle from child to child. The child holding the hat when the music stops puts the hat on.

Any prop that is quick and easy to put on can be used, such as toy spectacles, red nose, false moustache.

PERSONAL AND SOCIAL DEVELOPMENT

Talking Toy (i)

Aim
To be able to take turns at speaking.

Any toy, which will be named the 'Talking Toy' and used in all Talking Toy activities. A toy creature is a good idea.
Tape recorder and music tape, or piano.

The children sit in a semi-circle in front of you. Explain that the Talking Toy will be passed around the circle, and whoever is holding it when the music stops will say his name. Then the music will start again, and the toy continue on its way around the group.

Level I

Level II

Level III

Level IV

Circle Time

Hall/PE

Literacy

Topic Work

Drama

Small Group

© C Delamain &
J Spring 2000.
Photocopiable

PERSONAL AND SOCIAL DEVELOPMENT

- Level I
- Level II
- Level III
- Level IV
- **Circle Time**
- Hall/PE
- Literacy
- Topic Work
- Drama
- Small Group

Group Lotto

Aim
To be able to take turns in a cooperative activity.

Large picture lotto board and its matching pictures. A soft bag.

The small pictures are put in the bag, and the lotto board is put on the table. Explain that the children will take turns to take a picture out of the bag and place it correctly on the board. The bag is passed round the group and the children choose and place the pictures until the board is complete.

© C Delamain & J Spring 2000. Photocopiable

PERSONAL AND SOCIAL DEVELOPMENT

Balloon Bubbles (i)

Aim
To be able to take turns in a competitive game with adult support.

Boards and balloons from the commercial Match-a-Balloon® game.
A soft bag.

**A game for four to six children. Give a board to each child. Each board depicts six balloons or bubbles (red, blue, yellow, green, white and orange). The round balloon or bubble shapes (one set for each baseboard) are placed in the bag. The bag is passed round and the children take turns to draw out a bubble and place it on the matching shape on their baseboards.
When a child draws a colour that he already has the bubble is returned to the bag, so that the children begin to understand the concept of 'can't go' and 'already got that colour'. The die is *not* used.**

The equipment needed for this game can easily be made and does not have to come from a commercial source. You need a baseboard for each child depicting six bubbles of different colours, and six matching coloured bubble shapes for each board.

Level I

Level II

Level III

Level IV

Circle Time

Hall/PE

Literacy

Topic Work

Drama

Small Group

© C Delamain &
J Spring 2000.
Photocopiable

DEVELOPING BASELINE COMMUNICATION SKILLS

PERSONAL AND SOCIAL DEVELOPMENT

- Level I
- Level II
- Level III
- Level IV
- Circle Time
- Hall/PE
- Literacy
- Topic Work
- Drama
- Small Group

TURN TAKING

The Farmer Wants a Horse

Aim
To encourage children to take turns in choosing each other.

Equipment
None, unless the music is available.

How to Play
Children form a circle. One child is placed in the middle as the 'farmer'. The song is sung or chanted by you and the group, and the 'farmer' chooses other children in turn to join him in the middle as horse, goat, cat, etc. (See Activity Resources p228 for words of the song.)

© C Delamain & J Spring 2000. Photocopiable

PERSONAL AND SOCIAL DEVELOPMENT

Talking Toy (ii)

Aim
To take turns telling the group something you like.

Talking Toy.
Tape recorder and music tape, or piano.

Children sit in a semi-circle in front of you. Explain that when the music stops the child holding the toy must name one food that he likes. Talking Toy is passed along the line from child to child.

Other categories that you might choose: animals, toys, sports, television programmes, books.

Level I

Level II

Level III

Level IV

Circle Time

Hall/PE

Literacy

Topic Work

Drama

Small Group

© C Delamain &
J Spring 2000.
Photocopiable

DEVELOPING BASELINE COMMUNICATION SKILLS   PAGE 13

PERSONAL AND SOCIAL DEVELOPMENT

Pull Out a Name

Aim
To be able to allow other people a turn.

Slips of paper with the name of a child on each, one for every child.
A soft bag.
A large picture from a colouring book, preferably enlarged, and with plenty of small areas needing different colours.
Felt-tips or crayons.

The children sit in a circle around you. Put the names into the bag. Pin the picture up on the wall or stick it down on floor or table with Blu-Tack®. Let the first child pull a name out of the bag. Read it out, and the child gives it to the owner of the name who colours a piece of the picture of his own choosing. Then *he* pulls a name out of the bag and gives it to its owner to have a turn at colouring. Carry on round the group.

The idea of names in a bag and children selecting *other* children for turns can be applied to any joint project or activity.

© C Delamain &
J Spring 2000.
Photocopiable

PERSONAL AND SOCIAL DEVELOPMENT

Number's Up!

Aim
To be able to take turns to clap the number of dots on a die.

Giant die. If the group members are still at a very early stage of number concept and counting, use a cardboard cube, or brick, and draw or stick on dots going up to only three or four.

Children sit in a circle. The die is passed round the group, and when you say 'stop' the child holding the die throws it into the middle of the circle and claps out the number of dots. Then the die resumes its progress round the circle.

- Level I
- **Level II**
- Level III
- Level IV
- **Circle Time**
- Hall/PE
- Literacy
- Topic Work
- Drama
- **Small Group**

© C Delamain &
J Spring 2000.
Photocopiable

PERSONAL AND SOCIAL DEVELOPMENT

Add To It

Aim
To be able to wait for a given signal.

Giant building bricks, or giant stackers.

Children sit in a semi-circle in front of you. There is a pile of giant building bricks or stackers in the middle of the circle. Explain that this is a 'listen and wait' game. Choose a child and say ' . . ., when I say "Now" come and start the tower'. In turn, choose children to come and add another brick to the tower, each time making them wait after hearing their name until you say 'Now!'. The child whose brick causes the tower to fall down starts the next tower.

PERSONAL AND SOCIAL DEVELOPMENT

Nursery Rhyme Circle

Aim
To be able to take turns saying the words of any well-known nursery rhyme.

None.

Children sit in a semi-circle in front of you. Decide on a suitable nursery rhyme. This must either be one that the majority of children know really well, or one that has been taught and practised beforehand. The first child chants the first line of the rhyme, the second child carries it on by chanting the second line, and so on round the circle until the whole rhyme has been said. You may need to use a 'stop' signal such as holding up your hand, to indicate when a turn has finished.

Level I

Level II

Level III

Level IV

Circle Time

Hall/PE

Literacy

Topic Work

Drama

Small Group

© C Delamain &
J Spring 2000.
Photocopiable

DEVELOPING BASELINE COMMUNICATION SKILLS

PERSONAL AND SOCIAL DEVELOPMENT

TURN TAKING

Feely Bag

Aim
To be able to wait for an increasing length of time for your turn.

Equipment
Soft bag.
Assortment of objects, not too obvious in shape.

How to Play
Children sit in a semi-circle in front of you. One object is put into the bag without the children seeing it. The bag is passed round the circle, with every child having a feel. When the bag has gone around the whole circle, you say 'Hands up anyone who thinks they know what was in the bag'. One child is then chosen to make a guess, and guessing continues until somebody guesses correctly.

Tip
If this proves too difficult, start by showing the children an array of five or six objects. Let them feel each one, and only then put one of them secretly into the bag. Explain that the mystery object will be one of the objects they have already explored.

© C Delamain &
J Spring 2000.
Photocopiable

PERSONAL AND SOCIAL DEVELOPMENT

Yes-No

Aim
To learn that the order of turn-taking can vary, and to be able to give up a turn to someone else.

Slips of paper or card, enough for all members of the group. Half have 'Yes' written on them, and half have 'No'.
A soft bag.

This game can be played during any group activity such as making a floor puzzle, building a tower of giant bricks, or making a collage. Turns are regulated by the Yes/No bag. The bag is given to the first child, who pulls out a slip. If it is a 'Yes', that child takes his turn. If it is a 'No', the slip goes back in the bag, and the next child in the circle has a turn.

Level I

Level II

Level III

Level IV

Circle Time

Hall/PE

Literacy

Topic Work

Drama

Small Group

© C Delamain &
J Spring 2000.
Photocopiable

DEVELOPING BASELINE COMMUNICATION SKILLS

PERSONAL AND SOCIAL DEVELOPMENT

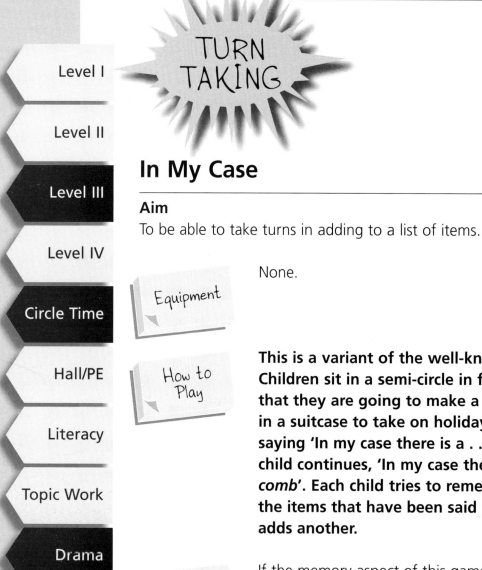

In My Case

Aim
To be able to take turns in adding to a list of items.

Equipment
None.

How to Play
This is a variant of the well-known game. Children sit in a semi-circle in front of you. Explain that they are going to make a list of things to go in a suitcase to take on holiday. You start by saying 'In my case there is a . . . *book*' The next child continues, 'In my case there is a *book* and a *comb*'. Each child tries to remember and repeat the items that have been said previously, and adds another.

Tip
If the memory aspect of this game is too hard for any child, let him just add his own item. If all the children find it too hard, again let them just name their own item. You can write the ideas down and read them back at the end.

Extension
The game can be varied by packing the case for a holiday in the sun, somewhere cold and snowy, a trip to the zoo, or to hospital etc.

© C Delamain &
J Spring 2000.
Photocopiable

PERSONAL AND SOCIAL DEVELOPMENT

Build It

Aim
To be able to cooperate in pairs in the building of a Lego®, Duplo® or Multilink® model.

Good supply of Lego®, Duplo® or Multilink®. For each pair, a coloured picture of a simple Lego®, Duplo® or Multilink® model (available in Activity Resources pp229–230).

Put children into pairs for this game. Give each child his own small supply of building equipment. You will need to ensure that every *pair* has enough bricks between them to complete their design. Distribute the design pictures among the pairs, and explain that they have to help each other make the model.

This looks complicated, but is well worth the effort!

Level I

Level II

Level III

Level IV

Circle Time

Hall/PE

Literacy

Topic Work

Drama

Small Group

© C Delamain & J Spring 2000. Photocopiable

DEVELOPING BASELINE COMMUNICATION SKILLS

PERSONAL AND SOCIAL DEVELOPMENT

Balloon Bubbles (ii)

Aim
To be able to take turns in a competitive game with one other child, without adult support, except for initial demonstration and explanation.

Boards and balloons from the commercial Match-a-Balloon® game.
A soft bag.

Give a board to each child. Each board depicts six balloons or bubbles (red, blue, yellow, green, white and orange). The round balloon or bubble shapes (one set for each child) are placed in the bag. The children take turns to draw out a bubble and place it on the matching shape on their baseboard. When a child draws a colour that he already has the bubble is returned to the bag. The die is *not* used. The winner is the first child to cover all his bubbles.

The equipment needed for this game can easily be made and does not have to come from a commercial source. You need a baseboard for each child depicting six bubbles of different colours, and six matching coloured bubble shapes for each board.

© C Delamain &
J Spring 2000.
Photocopiable

DEVELOPING BASELINE COMMUNICATION SKILLS

PERSONAL AND SOCIAL DEVELOPMENT

Talking Toy (iii)

Aim
To be able to say their full name and address.

Talking Toy.
Tape recorder and music tape, or piano.
A list of the children's names and addresses for the teacher in case prompting is needed.

Children sit in a semi-circle in front of you. Explain that when the music stops the child holding the toy must name say his full name and address. Talking Toy is passed along the line from child to child.

- Level I
- Level II
- Level III
- **Level IV**
- **Circle Time**
- **Hall/PE**
- Literacy
- Topic Work
- Drama
- **Small Group**

© C Delamain & J Spring 2000. Photocopiable

PERSONAL AND SOCIAL DEVELOPMENT

- Level I
- Level II
- Level III
- **Level IV**
- **Circle Time**
- Hall/PE
- Literacy
- Topic Work
- Drama
- **Small Group**

30 Seconds

Aim
To be able to maintain a topic for a set length of time.

A bag of familiar objects or toys as prompts.
Stop-watch or watch with second hand.

Children sit in a semi-circle in front of you. They take turns to select an object from the bag, and have to talk about it until you say 'Stop'. You will need to time them.

You may want to start with 15 or 20 seconds.
30 seconds is longer than you think! You can vary the length of time according to the ability of the child.

© C Delamain &
J Spring 2000.
Photocopiable

PERSONAL AND SOCIAL DEVELOPMENT

Balloon Bubbles (iii)

Aim
To be able to take turns in a competitive game in a small group, without adult support, except for initial demonstration and explanation.

 Boards and balloons from the commercial Match-a-Balloon® game.
A soft bag.

 **Give a board to each child. Each board depicts six balloons or bubbles (red, blue, yellow, green, white and orange) The round balloon or bubble shapes (one set for each child) are placed in the bag. The children take turns to draw out a bubble and place it on the matching shape on their baseboard. When a child draws a colour that he already has the bubble is returned to the bag. The die is *not* used. The winner is the first child to cover all his bubbles.**

 The equipment needed for this game can easily be made and does not have to come from a commercial source. You need a baseboard for each child depicting six bubbles of different colours, and six matching coloured bubble shapes for each board.

 Let the group play the Match-a-Balloon® game correctly using the coloured die. If home-made equipment is being used, the die can be made from a small toy brick with the faces coloured, or coloured stickers on each face.

Level I

Level II

Level III

Level IV

Circle Time

Hall/PE

Literacy

Topic Work

Drama

Small Group

© C Delamain &
J Spring 2000.
Photocopiable

PERSONAL AND SOCIAL DEVELOPMENT

TURN TAKING

Story Line

Aim
To be able to contribute in turn to a continuous story.

None.

Children sit in a semi-circle in front of you. You start a story off, and each child takes a turn to add the next part. You will probably need to hold up your hand to signal when a turn is over.

'Goldilocks and the Three Bears', 'Three Little Pigs', 'Very Hungry Caterpillar', 'Sly Fox and the Red Hen'. Ladybird *Well-Loved Tales Level I* are an excellent source of suitable stories.

Make sure all the children are familiar with a story before choosing it for this activity.

To make things harder, try a story that you and the children make up as you go along.

PERSONAL AND SOCIAL DEVELOPMENT

Post Box

Aim
To be able understand that your turn may not come for several days. To be able to take responsibility for checking whether it is your turn, and telling the teacher.

A slip of paper or card for each child with his name on it.
A home-made post box.

Position the post box in a suitable spot in the classroom and explain to the children that every day there will be one name in the box. You will ensure that a different name is put in every day. The children are told that they must look in the box every day. If the name inside is not theirs, they must put the slip back. If it *is* theirs, they must take it to you. It will then be their turn to do some favourite task, such as feeding a classroom pet.

This is easier to organise than it sounds.

Level I

Level II

Level III

Level IV

Circle Time

Hall/PE

Literacy

Topic Work

Drama

Small Group

© C Delamain &
J Spring 2000.
Photocopiable

Developing Baseline Communication Skills
PERSONAL AND SOCIAL DEVELOPMENT

Level I

30 / Watch Me! (i)
31 / Thumbs Up!
32 / Magic Messages!
33 / Magic Mime
34 / Magic Movements (i)

Level II

35 / Magic Box (i)
36 / One Thing for Another
37 / Elves and Goblins
38 / Follow My Leader
39 / Grandmother's Footsteps

Level III

40 / Magic Movements (ii)
41 / Gesture Sentences
42 / Statues
43 / Special Sitting
44 / Watch Me! (ii)

Level IV

45 / Fidget Fiends
46 / Mime Story
47 / Magic Movements (iii)
48 / Magic Box (ii)
49 / Hurrah-Boo!

PERSONAL AND SOCIAL DEVELOPMENT

- Level I
- Level II
- Level III
- Level IV
- **Circle Time**
- Hall/PE
- Literacy
- Topic Work
- Drama
- Small Group

Watch Me! (i)

Aim
To be able to pick up clues from watching another person.

None.

The children sit in a semi-circle in front of you. Explain that they are to respond by standing up, or raising a hand, when you look at them. You look at the children in random order. No speaking!

Let a child be the 'looker'.

© C Delamain &
J Spring 2000.
Photocopiable

PERSONAL AND SOCIAL DEVELOPMENT

Thumbs Up!

Aim
To be able to use a gesture to indicate 'yes' and 'no'.

None.

The children sit in a semi-circle in front of you. They are shown how to use the 'thumbs up' or 'thumbs down' signs for 'yes' and 'no'. You ask a question requiring a 'yes' or 'no' response. The children must not speak, only respond with gesture.

Is it sunny today?
Is my name . . . ?
Do people eat apples?
Do we drink milk?
Could we eat an elephant?
Have I got a hat on?

Level I

Level II

Level III

Level IV

Circle Time

Hall/PE

Literacy

Topic Work

Drama

Small Group

© C Delamain &
J Spring 2000.
Photocopiable

PERSONAL AND SOCIAL DEVELOPMENT

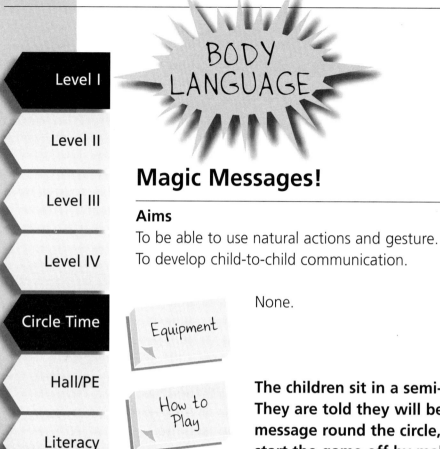

- Level I
- Level II
- Level III
- Level IV
- Circle Time
- Hall/PE
- Literacy
- Topic Work
- Drama
- Small Group

Magic Messages!

Aims
To be able to use natural actions and gesture.
To develop child-to-child communication.

Equipment

None.

How to Play

The children sit in a semi-circle in front of you. They are told they will be passing a magic message round the circle, without words. You start the game off by making a simple gesture to the nearest child. That child must repeat the gesture to his neighbour, and so on along the line.

Tip

Start with a really easy gesture, such as waving a hand, or touching your nose. Increase the difficulty by making the gesture more complex, eg, touch your nose *and then* your ear, wave your hand *and then* give a clap.

© C Delamain &
J Spring 2000.
Photocopiable

DEVELOPING BASELINE COMMUNICATION SKILLS

PERSONAL AND SOCIAL DEVELOPMENT

Magic Mime

Aim
To be able to use natural actions and gestures.

None.

The children sit in a semi-circle in front of you. They must guess what you are pretending to do. When you mime the actions, the children guess.

Brushing hair
Cleaning teeth
Waving
Going to sleep
Taking off/putting on shoes, coat, hat, gloves.
Eating something

Level I

Level II

Level III

Level IV

Circle Time

Hall/PE

Literacy

Topic Work

Drama

Small Group

© C Delamain &
J Spring 2000.
Photocopiable

PERSONAL AND SOCIAL DEVELOPMENT

Magic Movements (i)

Aims
To be able to mime the way animals move.

Equipment
Tape recorder and music tape or piano.

How to Play
The children stand in a clear space around the hall. They are told that when the music starts they are to move about like a certain animal. They are to stop when the music stops. You may need to demonstrate the movements at first. To start with the whole group mimes the same animal.

Examples
Bird
Snake
Elephant
Mouse
Frog
Cat

Extension
Let the children try to carry out the movements without a demonstration.
Give each child a different animal to mime.

PERSONAL AND SOCIAL DEVELOPMENT

Magic Box (i)

Aim
To be able to match the appropriate body language to an object.

A box.
An assortment of small objects or pictures. These should include some nice, some unpleasant, and some scary items.

Show the children the objects and invite discussion about how each item might make one feel. You then put one item into the box, unseen by the children. The box is passed to the first child, who must mime delight, horror, disgust or fear, and try to mime what the object is. The rest of the group must try to guess the object.

Spider, ice cream, trumpet, book, crocodile, piano, chocolate, spoonful of medicine, snake, cuddly toy.

Level I

Level II

Level III

Level IV

Circle Time

Hall/PE

Literacy

Topic Work

Drama

Small Group

© C Delamain &
J Spring 2000.
Photocopiable

PERSONAL AND SOCIAL DEVELOPMENT

- Level I
- **Level II**
- Level III
- Level IV
- **Circle Time**
- Hall/PE
- Literacy
- Topic Work
- Drama
- **Small Group**

One Thing for Another

Aims
To be able to pretend one object is another.

Equipment
An assortment of items from the classroom.
Slips of paper with actions written on them.
Hat or soft bag.

How to Play
Place the action slips in a hat or bag. The first child chooses one. Read the action to the child in a whisper, and hand him one of the classroom objects (eg, a pencil). He must then mime the action using the pencil as a prop. The rest of the group guess what he is miming.

Examples of Objects
Pencil, ruler, gluepot, pencil case, rubber.

Examples of Actions
Telephoning, ironing, writing, powdering your nose, putting on lipstick, brushing hair, eating soup.

© C Delamain & J Spring 2000. Photocopiable

PERSONAL AND SOCIAL DEVELOPMENT

Elves and Goblins

Aims
To be able to understand and respond to welcoming and rejecting gestures.

Sticky labels.
Two large pieces of card, one with a red triangle on it and the other with a black square.

One child (A) is chosen as elf leader, and one child (B) as goblin leader. They sit or stand in opposite corners of the room. A has the red triangle in front of him on the floor, B has the black square. The rest of the class have sticky labels applied to their backs, half with red triangles and half with black squares. They go from corner to corner, displaying their backs to the elf and goblin leaders. Leaders beckon to members of their clan to come and join them, but wave members of the other clan away. The game ends when all the children have assembled in their correct corner.

Have an elf, goblin, witch and wizard corner.

Level I

Level II

Level III

Level IV

Circle Time

Hall/PE

Literacy

Topic Work

Drama

Small Group

© C Delamain & J Spring 2000.
Photocopiable

DEVELOPING BASELINE COMMUNICATION SKILLS

PERSONAL AND SOCIAL DEVELOPMENT

Follow My Leader

Aim
To be able to watch others' movements carefully and copy them.

None.

Choose a child as leader. He sets off round the room with a particular action. One by one the rest of the children fall in behind him, copying the action until the whole line is doing it.

Walking, marching, hopping, skipping, jumping, crawling, walking backwards.

The leader can change the action as he goes along.

PERSONAL AND SOCIAL DEVELOPMENT

Grandmother's Footsteps

Aim

To be able to tiptoe quietly and slowly for long enough to reach the 'grandmother'.

None.

A well-known old-fashioned game. To start with, you should be grandmother. Stand at one end of the room with your back to the class. The children line up side by side some distance behind the grandmother. At a given signal they begin to creep slowly and silently forward. If grandmother hears a sound she turns round suddenly, and any child caught moving must go back to the starting line and begin again. Every so often, whether she hears a sound or not, grandmother turns round swiftly and tries to catch someone moving. The winner is the child who can reach grandmother and tap her on the shoulder without getting caught.

Let a child be the 'grandmother'.

© C Delamain & J Spring 2000.
Photocopiable

DEVELOPING BASELINE COMMUNICATION SKILLS

PERSONAL AND SOCIAL DEVELOPMENT

Magic Movements (ii)

Aim
To be able to mime the ways in which other people move.

None.

The children stand spread out around the hall. Explain that they are to move around the hall in the way that some other people might move.

An old man with a stick
A toddler who can only just walk
A mother carrying heavy shopping
A lady pushing a pram
A runner
Someone with a bad leg
A dancer

Choose one child and let him decide for himself which of the movements they have practised he would like to do. He mimes the movement, the rest of the group guess which he is doing.

PERSONAL AND SOCIAL DEVELOPMENT

Gesture Sentences

Aim
To be able to convey a simple message using gesture.

None.

The children sit in a semi-circle in front of you. Choose a child and whisper a simple message to him. Explain that he is to show this message to the group without talking. The rest of the children guess the message.

Come here!
Go away!
Sit down!
Turn round!
He can smell a horrible smell
He's too hot or too cold
He's tasting something delicious
He has a tummy ache

PERSONAL AND SOCIAL DEVELOPMENT

- Level I
- Level II
- **Level III**
- Level IV
- Circle Time
- **Hall/PE**
- Literacy
- Topic Work
- **Drama**
- Small Group

Statues

Aim
To be able to stand still.

None.

The children should spread out around the room. One child is chosen to be the 'wizard' who can turn them into stone. He goes around the room touching each child lightly on the shoulder. As he touches, the child must 'freeze' into a statue. Nobody must move until the wizard comes round again to touch the statues and bring them back to life.

PERSONAL AND SOCIAL DEVELOPMENT

Special Sitting

Aim
To be able to differentiate between 'good' and 'bad' sitting positions.

Large teddy or doll.

The children sit in semi-circle in front of you. You have a spare chair next to you, on which you put the teddy or doll. Explain that the teddy is a very bad sitter. Using the teddy, demonstrate lolling, looking away, wriggling, fidgeting with hands. Then use the teddy to demonstrate good sitting, back straight, feet on the floor, looking at teacher, hands still on lap.

Tell the children they are going to practise good and bad sitting. Go round the group touching each child and instructing them to do good sitting or bad sitting. At the end, tell them that when you clap your hands *everyone* is to do good sitting while you count slowly to 10.

- Level I
- Level II
- **Level III**
- Level IV
- **Circle Time**
- Hall/PE
- Literacy
- Topic Work
- Drama
- **Small Group**

© C Delamain &
J Spring 2000.
Photocopiable

PERSONAL AND SOCIAL DEVELOPMENT

- Level I
- Level II
- **Level III**
- Level IV
- **Circle Time**
- Hall/PE
- Literacy
- Topic Work
- Drama
- **Small Group**

Watch Me! (ii)

Aim
To be able to copy a series of movements made by you.

None.

The children sit in a semi-circle in front of you. Explain that the children must do exactly what you do. Carry out several series of different movements, changing from one movement to another increasingly fast. A second adult is needed to spot any child who falls behind or confuses the movements. They are told they are 'out' and stop copying the movements. The last child in is the winner.

Wave one hand, wave both hands, tap your nose, tap one knee with one hand, tap both knees, pat your tummy with hands and so on.

© C Delamain &
J Spring 2000.
Photocopiable

PERSONAL AND SOCIAL DEVELOPMENT

Fidget Fiends

Aim

The children will resist the temptation to fidget for up to five minutes.

An assortment of small toys.

The children sit in a semi-circle in front of you. A little 'fidgety' toy is put in front of each child. Explain that you are going to read a short story and that nobody is to touch the toys until the story is over. As any child gives in to temptation and picks up the toy, that toy goes in to a 'fidget' box in the middle. The winner is the child whose toy is still in front of him when the story ends.

- Level I
- Level II
- Level III
- **Level IV**
- **Circle Time**
- Hall/PE
- Literacy
- Topic Work
- Drama
- **Small Group**

© C Delamain &
J Spring 2000.
Photocopiable

PERSONAL AND SOCIAL DEVELOPMENT

Mime Story

Aim
To be able to mime suitable actions and expressions to accompany a story read aloud.

Equipment

Story texts (in Activity Resources pp231–232).

How to Play

Explain that you will read a story aloud. The children are to mime the actions as the story goes along.

Tip

You may have to begin very slowly and prompt each action until the children get into the swing of things. The story is written with ( ) marks to indicate where to pause, and to prompt if necessary.

© C Delamain & J Spring 2000. Photocopiable

PERSONAL AND SOCIAL DEVELOPMENT

Magic Movements (iii)

Aim
To be able to mime movements and actions suitable for someone with a specific occupation.

None.

The children spread out around the room. You call out an occupation. The children must interpret how they might feel and move if they were engaged in that activity, and mime appropriately.

Postman	Doctor
Dentist	Hairdresser
Teacher	Builder
Milkman	Checkout clerk
Fisherman	Footballer
Lorry driver	Window cleaner

Level I

Level II

Level III

Level IV

Circle Time

Hall/PE

Literacy

Topic Work

Drama

Small Group

© C Delamain &
J Spring 2000.
Photocopiable

DEVELOPING BASELINE COMMUNICATION SKILLS

PERSONAL AND SOCIAL DEVELOPMENT

BODY LANGUAGE

Magic Box (ii)

Aim
To be able to mime an imaginary object.

A box.

The box is passed to the first child. Explain that he is to look in the box and pretend there is something in it. He must try to convey to the class without words what it is he is pretending to see. The rest of the class guess. The next child has a turn, and so on round the group.

You may need to offer suggestions at first as to what might be in the box.

© C Delamain &
J Spring 2000.
Photocopiable

DEVELOPING BASELINE COMMUNICATION SKILLS

PERSONAL AND SOCIAL DEVELOPMENT

Hurrah-Boo!

Aim
To be able to react with actions to rapidly changing ideas.

None.

The children spread out with plenty of room between them, but within easy listening distance of you. Explain that whenever you say something pleasant the children are to react by shouting 'Hurrah!' and raising their arms in the air. If you say something unpleasant, they are to shout 'Boo!' and bend down towards the floor. Start slowly and speed the changes up gradually. Randomise the prompts, so that the children really have to listen. Do not just alternate pleasant/unpleasant, but perhaps two pleasant, one unpleasant, one pleasant, two or three unpleasant.

We're going swimming
It's raining and we'll have to stay in
We're going on a trip to the zoo
The picnic has been cancelled
We're all going to get bad colds
Father Christmas is coming to visit
The holidays have been cancelled
There will be a party next week

This activity is a good warm-up, and also useful for letting off steam if the children are 'high' or over-excited!

- Level I
- Level II
- Level III
- **Level IV**
- Circle Time
- **Hall/PE**
- Literacy
- Topic Work
- Drama
- Small Group

© C Delamain &
J Spring 2000.
Photocopiable

DEVELOPING BASELINE COMMUNICATION SKILLS

Developing Baseline Communication Skills
PERSONAL AND SOCIAL DEVELOPMENT

AWARENESS OF OTHERS

Level I

52 / Empty Chair
53 / Who Has Gone?
54 / About Us
55 / Who is Asleep?
56 / Listen and Jump!

Level II

57 / Lining Up
58 / Ask Me!
59 / Find Me!
60 / Yummie Yuckie
61 / Squashed Bananas

Level III

62 / Guess Who?
63 / News
64 / Alien Visitor (i)
65 / Four Corners of the Earth
66 / New Kid!

Level IV

67 / Alien Visitor (ii)
68 / Help!
69 / Gift Box
70 / Wash the Puppy
71 / Who Am I?

PERSONAL AND SOCIAL DEVELOPMENT

Empty Chair

Aim
To learn and be able to use each others' names.

None.

Children sit on chairs in a semi-circle in front of you. There is one extra chair that is empty. Say 'Oh look, there is an empty chair, 'A', [naming a child on one side of the empty chair], who would you like to come and sit by you?' Child A is encouraged to choose someone. This may be only by pointing at first. You supply the name of the chosen child (child B) if necessary, and help child A to invite him – 'B, please come and sit here'. Child B changes place. You then feign great surprise – 'Now there is *another* empty chair!' Repeat with the next child.

© C Delamain &
J Spring 2000.
Photocopiable

DEVELOPING BASELINE COMMUNICATION SKILLS

PERSONAL AND SOCIAL DEVELOPMENT

Who Has Gone?

Aim
To learn each others' names.

None.

The children sit on chairs in a semi-circle in front of you. The children are told to close their eyes. Choose one child to leave the group and go out of sight. (This may need a second adult's help at first, particularly if the child needs to leave the room in order to be out of sight.) The children are told to open their eyes, and you ask 'Who has gone?' The children try to remember who was sitting on the chair that is now empty. Repeat with the rest of the children.

Level I

Level II

Level III

Level IV

Circle Time

Hall/PE

Literacy

Topic Work

Drama

Small Group

© C Delamain &
J Spring 2000.
Photocopiable

PERSONAL AND SOCIAL DEVELOPMENT

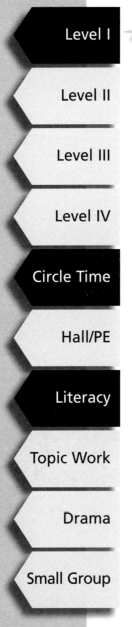

About Us

Aim
To be able to recognise characteristics of others.

None.

The children sit in a semi-circle in front of you. Choose a child (A), and ask a question about his next-door neighbour (child B) such as 'What colour jumper is B wearing' (or 'Is he wearing a tie', or 'Has he got trainers on?'). Child A must look carefully at his neighbour, (B) and respond. Then it is B's turn to be asked a question about *his* neighbour (child C). Continue in this way round the circle.

Questions can also be tailored for children who do not yet know their colours.

© C Delamain &
J Spring 2000.
Photocopiable

PERSONAL AND SOCIAL DEVELOPMENT

Who is Asleep?

Aim
To be able to look carefully at each others' faces.

A soft bag.
Small home-made cards, as many as there are children in the group. One card shows two eyes closed, the rest show eyes open (a template for open and closed eyes is available in Activity Resources p233).

The children sit in a semi-circle facing you. Give each child a turn to take a card out of the bag, hidden from the group. Explain that the child who draws the 'eyes closed' card will close his eyes and pretend to be asleep when you say 'NOW!'. The rest of the class must look carefully around at each other to see who is 'asleep'. Put cards back in the bag and repeat.

- Level I
- Level II
- Level III
- Level IV
- Circle Time
- Hall/PE
- Literacy
- Topic Work
- Drama
- Small Group

© C Delamain &
J Spring 2000.
Photocopiable

PERSONAL AND SOCIAL DEVELOPMENT

Listen and Jump!

Aim
To develop awareness of own appearance.

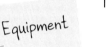

None.

The children stand in a circle around you. Give a simple command involving physical appearance. Explain to the children that if what you say applies to them, they must jump into the middle of the circle.

Encourage discussion if a child fails to jump when he should. Do the others think he *does* have brown eyes/black hair/freckles?

Easy 'If you've got *trainers* on, jump'
(recognition of noun)

Harder – 'If you've got *black shoes/brown hair/long trousers* on, jump'
(understanding of noun + adjective)

© C Delamain &
J Spring 2000.
Photocopiable

PERSONAL AND SOCIAL DEVELOPMENT

Lining Up

Aim
To be able to recognise own and others' characteristics.
To develop child-to-child talk.

None.

Choose a leader, who should be the tallest child in the class or group. The leader must choose the child he thinks is closest to himself in height, and ask that child to come and stand next to him. The second child then chooses the next tallest again, and so on until a complete line is formed of the whole class or group. Asking children to stand back to back to compare heights is to be encouraged. At the end a child can be chosen to study the line, and re-arrange it where he thinks mistakes have been made.

Level I

Level II

Level III

Level IV

Circle Time

Hall/PE

Literacy

Topic Work

Drama

Small Group

© C Delamain &
J Spring 2000.
Photocopiable

PERSONAL AND SOCIAL DEVELOPMENT

- Level I
- Level II
- Level III
- Level IV
- Circle Time
- Hall/PE
- Literacy
- Topic Work
- Drama
- Small Group

Ask Me!

Aim
To be able to ask a question in order to find out some specific information. To develop child-to-child talk.

None.

The children sit in a semi-circle around you. Choose a simple question, for example, 'Have you got any pets at home?' Child A is instructed to ask this question of his neighbour, (child B) who relays the answer to the whole group. Then B asks *his* neighbour, who replies, and so on round the circle.

The amount of elaboration allowed on the replies given will be at your discretion.

© C Delamain & J Spring 2000. Photocopiable

PERSONAL AND SOCIAL DEVELOPMENT

Find Me!

Aim
To be able to recognise own and each others' physical characteristics.

Set of home-made cards showing eyes of different colours or long/short hair of different colours.

The children are given one card each. They move around the room and try to find someone who matches the characteristics on their card. When they think they have found somebody suitable, they go and tell the teacher.

Level I

Level II

Level III

Level IV

Circle Time

Hall/PE

Literacy

Topic Work

Drama

Small Group

© C Delamain &
J Spring 2000.
Photocopiable

PERSONAL AND SOCIAL DEVELOPMENT

- Level I
- **Level II**
- Level III
- Level IV
- Circle Time
- Hall/PE
- Literacy
- Topic Work
- **Drama**
- **Small Group**

Yummie Yuckie

Aim
To be able to express own likes and dislikes, and listen to others' likes and dislikes.

Two identical sets of coloured counters or small bricks, enough for each child to have a different coloured counter or brick.

An activity for a maximum of six children. The children sit in a circle. Give each child a brick or counter. Put the duplicate set out of sight, in a container. Tell the children to put their brick under their chair. Explain that everyone is going to have a turn saying a food they like. Go round the circle, letting each child contribute a favourite food.

Now choose a child (A) to take a brick or counter out of the container. Ensure that it is a different colour from his own. Identify who has the matching brick (child B) and point him out. Child A has to try to remember what child B chose as his favourite food. Play continues until everyone has had a turn at remembering.

PERSONAL AND SOCIAL DEVELOPMENT

Squashed Bananas

Aim
To be aware of what makes other children laugh.

None.

The children sit in a semi-circle facing you. One child (A) is chosen to answer questions from the rest of the group. Explain that the only answer they can give is 'squashed bananas'. The rest of the group take turns to ask questions to try to make A laugh. As soon as A laughs, another child takes his place.

What do you wear in bed?
What does mum wash her hair with?
What have you got in your pocket?

You may have to help by whispering question suggestions.

- Level I
- **Level II**
- Level III
- Level IV
- **Circle Time**
- Hall/PE
- Literacy
- Topic Work
- Drama
- **Small Group**

© C Delamain &
J Spring 2000.
Photocopiable

PERSONAL AND SOCIAL DEVELOPMENT

Guess Who?

Aim
To be able to identify someone from a description.

None.

Start to describe a child in the group. 'It's a girl. She has long brown hair and brown eyes. She is wearing a blue skirt and a white shirt. She has gold ear studs …' As soon as someone thinks they have guessed who it is, they must put their hand up. Then repeat the process with another child.

© C Delamain &
J Spring 2000.
Photocopiable

PERSONAL AND SOCIAL DEVELOPMENT

News

Aim

To be able to listen to what others say and retell to the rest of the group.

None.

The children sit in a semi-circle facing you. Each child gives an item of news, from the previous weekend or day. Then each child tells the group their right-hand neighbour's news.

Level I

Level II

Level III

Level IV

Circle Time

Hall/PE

Literacy

Topic Work

Drama

Small Group

© C Delamain &
J Spring 2000.
Photocopiable

PERSONAL AND SOCIAL DEVELOPMENT

Alien Visitor (i)

Aim
To be aware of how to treat a visitor.

 A doll or creature to represent the 'alien visitor'.

 **The children sit in a semi-circle facing you. Introduce the 'doll' to the children as a visitor called Zig, from an alien planet. Explain that Zig feels nervous and worried. The 'alien visitor' is passed round the circle and each child is asked to say something to make him feel welcome before passing him on to the next child.**

 You may need to prompt, with ideas of suitable greetings.

'Hello Zig.'
'Come and play with us.'
'Sit next to me.'

 Give Zig to two children to look after for the rest of the day.

DEVELOPING BASELINE COMMUNICATION SKILLS

PERSONAL AND SOCIAL DEVELOPMENT

Four Corners of the Earth

Aim
To be able to recognise own and others' likes and dislikes.

None.

The teacher chooses a familiar category and selects four items in the category. A corner of the room is assigned to each item. Group the children in the middle of the room and explain that you are going to call out one of the items. If it is their favourite in that category, they go to that corner of the room.

Pets, toys, colours.

It may be necessary to put one picture of the category items (eg, dog, cat, horse, etc) in each corner to help the children understand where to go.

- Level I
- Level II
- **Level III**
- Level IV
- Circle Time
- Hall/PE
- Literacy
- Topic Work
- **Drama**
- **Small Group**

© C Delamain &
J Spring 2000.
Photocopiable

PERSONAL AND SOCIAL DEVELOPMENT

- Level I
- Level II
- **Level III**
- Level IV
- **Circle Time**
- Hall/PE
- Literacy
- Topic Work
- Drama
- **Small Group**

AWARENESS OF OTHERS

New Kid!

Aim
To be aware of how it feels to be a 'new' pupil, and recognise what the new pupil would need to know.

Picture or photograph of an unknown child.

Tell the class that a new pupil will be joining them today. Explain that this pupil has just moved here from a long way away. Each member of the group then has the chance to tell the 'new kid' something about the school, routine, etc, that will be useful.

© C Delamain &
J Spring 2000.
Photocopiable

PERSONAL AND SOCIAL DEVELOPMENT

Alien Visitor (ii)

Aim
To be aware of the environment, in the classroom and in the playground.

'Zig', the alien visitor as used in Alien Visitor (i).

Explain that several of the alien's friends are coming to visit the school. Ask the children how they can make the classroom or the playground look pleasing for the visitors. The children are encouraged to offer ideas, and then help to tidy up, remove litter, decorate the classroom etc.

This activity could be incorporated into topic work about the immediate environment, and form the basis of simple writing tasks, art work, etc.

- Level I
- Level II
- Level III
- Level IV
- Circle Time
- Hall/PE
- Literacy
- Topic Work
- Drama
- Small Group

© C Delamain &
J Spring 2000.
Photocopiable

DEVELOPING BASELINE COMMUNICATION SKILLS

PERSONAL AND SOCIAL DEVELOPMENT

- Level I
- Level II
- Level III
- **Level IV**
- **Circle Time**
- Hall/PE
- Literacy
- Topic Work
- Drama
- **Small Group**

Help!

Aim
To help children identify and solve everyday problems.

A list of 'problem' scenarios, written on a set of cards and put in a container.

Explain to the group that you are having a 'bad day', and need their help sorting out a few problems. Choose a child (A) to select a card for you. Then, in a convincing way, you tell them what has happened – a bit of melodrama will add interest here!

Ask if anyone has a good solution to the problem, collecting about five or six solutions, and then try to decide which would be the best. This activity may take quite a long time, in which case you can choose just one scenario per session.

Left lunchbox at home.
Forgot PE kit.
Don't understand what to do.

© C Delamain &
J Spring 2000.
Photocopiable

PERSONAL AND SOCIAL DEVELOPMENT

Gift Box

Aim
To be able to select an appropriate gift based on knowledge of other people.

Cards with names of all members of group, and a container.

The children sit in a semi-circle facing you. One child (A), chooses a name out of the 'hat'. Child A then decides on a suitable present for that child, based on what he already knows about that child. Everyone has a turn.

Level I

Level II

Level III

Level IV

Circle Time

Hall/PE

Literacy

Topic Work

Drama

Small Group

© C Delamain &
J Spring 2000.
Photocopiable

DEVELOPING BASELINE COMMUNICATION SKILLS

PERSONAL AND SOCIAL DEVELOPMENT

Wash the Puppy

Aim
To be able to work cooperatively with a small group.

None, but a fairly large space should be available.

The children should be divided into groups of approximately four. Explain that your puppy has been playing in some puddles and is very dirty. Ask the groups if they would wash the puppy, reminding them to be gentle with him. Then each group acts out their version, with the rest watching.

Other scenarios needing good group cooperation:
Carrying a long plank of wood.
Putting up a tent.
Skipping with a long rope.
Passing a heavy log along the line.

PERSONAL AND SOCIAL DEVELOPMENT

Who Am I?

Aim
To be able to recognise a familiar person from the sound of their voice.

None.

The children stand in a line, one behind the other. The front child (A) puts on a blindfold and walks forward a few steps. The rest of the line silently change places. The second child (B) walks up to A and taps him gently on the shoulder, saying 'Who am I?' in a disguised voice. A tries to guess, and if wrong, asks simple questions which require 'yes' or 'no' as an answer, until B is identified. It is then B's turn, and so on.

You can suggest suitable questions, especially at the beginning of the activity.

- Level I
- Level II
- Level III
- **Level IV**
- Circle Time
- **Hall/PE**
- Literacy
- Topic Work
- **Drama**
- Small Group

© C Delamain &
J Spring 2000.
Photocopiable

DEVELOPING BASELINE COMMUNICATION SKILLS — PAGE 71

Developing Baseline Communication Skills
PERSONAL AND SOCIAL DEVELOPMENT

CONFIDENCE AND INDEPENDENCE

Level I

74 / Roundabout
75 / Take Rabbit
76 / Fetch It!
77 / Choosing Chain
78 / What You Need

Level II

79 / Pass the Beanbag
80 / Hot and Cold
81 / Sergeant Major
82 / 1-2-3 Choose!
83 / Auntie Jean's Birthday

Level III

84 / Lions and Tigers
85 / Scavenge Hunt
86 / Collections
87 / Specially Me
88 / Messengers

Level IV

89 / Job to Job
90 / Egg Timer
91 / Relay Race
92 / Tower of Babel
93 / Put the Tail on the Donkey

PERSONAL AND SOCIAL DEVELOPMENT

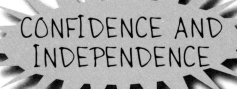

Roundabout

Aim
To be able to move independently within the group.

None.

Everyone sits in a circle. Name two children and they must stand up, step to the outside of the circle and run round in opposite directions and return to their own chairs. There must be enough space for children to get out of the circle and run round it. This continues until everyone has had a turn.

You can make the activity harder by telling the two children to run to *each other's* chairs instead of returning to their own.

PERSONAL AND SOCIAL DEVELOPMENT

CONFIDENCE AND INDEPENDENCE

Take Rabbit

Aim
To be able to move independently around the class.

Equipment

One toy rabbit, or similar.

How to Play

The children sit in a semi-circle facing you. Explain to the children that the rabbit wants to explore the classroom, but as he's only a toy he cannot move around on his own. Choose child A, and tell him where rabbit wants to go. Child A carries out the command, returns to the group and passes rabbit to B, and so on, until everyone has had a go.

Examples

'Take rabbit to the door'
'Show rabbit the window'
'Show rabbit where the book corner is'
'Rabbit wants to see the home corner'
'Take rabbit to the sink'

Level I

Level II

Level III

Level IV

Circle Time

Hall/PE

Literacy

Topic Work

Drama

Small Group

© C Delamain &
J Spring 2000.
Photocopiable

DEVELOPING BASELINE COMMUNICATION SKILLS

PERSONAL AND SOCIAL DEVELOPMENT

Fetch It!

Aim
To be able to instruct another child to fetch an object.

Equipment

A selection of objects: a pencil, book, shoe, paint brush, box, rubber, brick, ball etc. There should be at least as many objects as there are children, but some can be repeated, eg, several different coloured pencils.

How to Play

Divide the children into two teams: A and B. Each team stands in a line, one child behind the other. Put the objects on the floor at the other end of the room. The first child in team A tells the first child in team B which object to get. The child from team B brings it back, gives it to you and both children go to the back of the line. Continue until everyone has had a turn.

Tip

Remember which children have had turns at 'telling', so that next time you swap round to give the others an opportunity to tell.

© C Delamain &
J Spring 2000.
Photocopiable

PERSONAL AND SOCIAL DEVELOPMENT

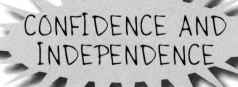

Choosing Chain

Aim
To be able to choose one child from the group.

None.

Explain to the children that you are going to make a line. One child is chosen to stand behind you. That child chooses someone to stand behind him, who chooses someone to stand behind him, etc, until the whole group is standing in a line. You then reverse the activity by inviting the last child to step forward and begin a new line by choosing someone.

Level I

Level II

Level III

Level IV

Circle Time

Hall/PE

Literacy

Topic Work

Drama

Small Group

© C Delamain &
J Spring 2000.
Photocopiable

PERSONAL AND SOCIAL DEVELOPMENT

CONFIDENCE AND INDEPENDENCE

What You Need

Aim
To be able to collect the right items to carry out a task.

A list of familiar classroom tasks.

The children sit in a semi-circle facing you. Explain that they will need to go and get the things they need for doing different tasks in the classroom. Tell child (A) one task to collect things for. The rest of the group can decide whether A has brought the right equipment. Then move on to the next child and so on until everyone has had a turn.

You can make the activity harder by giving each child an instruction, and telling them to wait until everyone has their instructions before collecting their equipment. This places the biggest memory load on the first child. You need to have a small group (maximum of six) for this to be successful, otherwise the memory load is too great and the children become restless.

Painting a picture
Making a birthday card
Writing a story
Sticking shapes in a maths book
Shared reading
Snakes and Ladders
Counting with bricks and counters
Colouring in a picture

© C Delamain &
J Spring 2000.
Photocopiable

PAGE 78   DEVELOPING BASELINE COMMUNICATION SKILLS

PERSONAL AND SOCIAL DEVELOPMENT

Pass the Beanbag

Aim

To be able to move round the group so that the game can continue.

Two beanbags.

This is a variation on a popular party game. Divide the children into two teams. Each team stands in a line, one behind the other. The front child in each team is handed a beanbag. When you say 'Go!' the front child in each team passes the beanbag over his head to the child behind, who passes it to the child behind him. When the last child in the team has the beanbag, he runs to the front, and that team is the winner.

You can vary the way in which the beanbag is passed, according to the age and ability of the children.

- Level I
- Level II
- Level III
- Level IV
- Circle Time
- Hall/PE
- Literacy
- Topic Work
- Drama
- Small Group

© C Delamain &
J Spring 2000.
Photocopiable

PERSONAL AND SOCIAL DEVELOPMENT

CONFIDENCE AND INDEPENDENCE

Hot and Cold

Aim
To be able to move independently round the class.

A small object which can be hidden.

Choose two children to be the hunters. They must hide their eyes while you hide the object somewhere in the classroom. If possible, have another adult who can distract the hunters and prevent cheating! Explain to the rest of the group that they must not tell the hunters where the object is, but that they are allowed to say things like 'hot', 'cold', 'a bit warmer', etc, according to how near to the object the hunters are. When the object has been found this is repeated with two new hunters.

Choose a single hunter, which requires more confidence and independence.

PERSONAL AND SOCIAL DEVELOPMENT

Sergeant Major

Aim
To be able to give a short message to a small group.

None.

The children are divided into two teams. Each team stands in a line facing the other team. A child from team A is the 'sergeant major'. This child gives a command to the other team, who have to do what they are told. The teams then swap and a child from team B is chosen to act as 'sergeant major'.

You may need to help children think of commands, especially at first.

Clap your hands
Turn around
Count to three
Touch the floor
Kneel down
Look up at the ceiling
Stand on one leg
Say 'cheese and biscuits'
Click your fingers
Sit on the floor

Level I

Level II

Level III

Level IV

Circle Time

Hall/PE

Literacy

Topic Work

Drama

Small Group

© C Delamain &
J Spring 2000.
Photocopiable

DEVELOPING BASELINE COMMUNICATION SKILLS

PERSONAL AND SOCIAL DEVELOPMENT

CONFIDENCE AND INDEPENDENCE

1-2-3 Choose!

Aim
To be able to make a choice quickly.

Equipment
A red and a blue counter for each child.

How to Play
The children stand in a semi-circle facing you. Explain that the red counter means 'sit on the floor', and the blue counter means 'clap your hands'. They will need to rehearse this a few times, so that they remember. Then each child puts the counters on the floor in front of them. Tell them you are going to count to three, and when you say 'THREE!' they must choose either red or blue, pick it up and do the activity that goes with that counter.

Extension
When they are used to doing the activity you can change the actions, eg, so that the red counter means 'turn around' and blue means 'stand on one foot' etc.

© C Delamain &
J Spring 2000.
Photocopiable

PERSONAL AND SOCIAL DEVELOPMENT

Auntie Jean's Birthday

Aim
To be able to think of necessary equipment for different activities.

A list of familiar activities to refer to.

Everyone sits in a circle. One child starts the game. Tell them what the activity is, and that child thinks of one item needed. The next child thinks of another item, and so on, until they have exhausted all the possibilities. Then a different activity is chosen and continue as before.

It's Auntie Jean's birthday
Going on a picnic
Feeding the classroom pet
Making biscuits
Making a packed lunch
Going to the beach
Decorating a Christmas tree
Swimming
Playing rounders
Washing the dishes

Level I
Level II
Level III
Level IV
Circle Time
Hall/PE
Literacy
Topic Work
Drama
Small Group

© C Delamain &
J Spring 2000.
Photocopiable

PERSONAL AND SOCIAL DEVELOPMENT

CONFIDENCE AND INDEPENDENCE

- Level I
- Level II
- **Level III**
- Level IV
- Circle Time
- **Hall/PE**
- Literacy
- Topic Work
- **Drama**
- Small Group

Lions and Tigers

Aim
To be able to organise the members of a group.

One set of pictures of lions in different positions, enough for half the class, and one set of tigers in different positions, enough for the remainder. (photocopiable pictures available in Activity Resources pp234–235).

The children are separated into two groups, lions and tigers, who stand in the middle of the room. A lion leader (child A) and a tiger leader (child B) are chosen by you. You have a pile of lion pictures and a pile of tiger pictures in front of you. Explain that the leaders are going to tell their animals where to go and what position to adopt. One corner of the room belongs to each group. Child A and child B come to you, and are given a picture from their pile. They must go and fetch a child from their group, show him where to go, and tell him whether he has to lie down, curl up, sit down or stand up, according to the picture he has been given. When the child is in the right position, the leader comes back to you, hands in his picture and collects a second one. The first group to be organised into the correct positions is the winning team.

© C Delamain &
J Spring 2000.
Photocopiable

DEVELOPING BASELINE COMMUNICATION SKILLS

PERSONAL AND SOCIAL DEVELOPMENT

Scavenge Hunt

Aim
To be able to move around the classroom independently.

A pictorial list of objects, one list per child. (Photocopiable pictures are available in Activity Resources, p236.)
A bag or container per child

The children sit in a semi-circle facing you. Explain that they are all going to have to find certain things in the classroom. Each child is given a list, and if necessary go through the items on the list so that they all know what to look for. When you say 'Go!' they start collecting items. The first child to collect them all is the winner.

Level I

Level II

Level III

Level IV

Circle Time

Hall/PE

Literacy

Topic Work

Drama

Small Group

© C Delamain & J Spring 2000. Photocopiable

PERSONAL AND SOCIAL DEVELOPMENT

CONFIDENCE AND INDEPENDENCE

Collections

Aim
To be able to move around the school grounds confidently.

Shopping bags or other plastic bags, enough for half the group.

The children are divided into pairs. Each pair is given a bag and a collecting target. What this is depends on the nature of the school grounds and the time of year. Targets might be to collect three different leaves, three different kinds of stone, three twigs. Where the school has no grass or trees it may be necessary to 'salt' the area first, by distributing shells, marbles, leaves, stones, counters etc.

It may be a good idea to pair timid children with more confident ones at first, but if the game is played more than once, two less confident children should be paired. If necessary, remind the children of the dangers of putting plastic bags over heads.

© C Delamain & J Spring 2000. Photocopiable

PERSONAL AND SOCIAL DEVELOPMENT

Specially Me

Aim
To enable the children to recognise ways in which they are individual and unique, and feel proud of them.

Pencil and paper for you.
Pictures of a collection of toys, a collection of clothes, a group of people, a collection of pets.

The children sit in a semi-circle in front of you. The pictures are placed on the floor in the middle of the circle. In turn the children are asked to say something that is special about themselves. You can use the pictures as prompts. 'Can you tell us about anyone in your family? About any pets? About any special clothes you like wearing? About any special toy you've got?' Write down the responses. When everyone has had a turn, read out the responses one by one. Ask the group who else would like some of the items on the list: a gerbil, some trendy trainers, a new baby sister . . . Hands up!

Use no prompts, but encourage the children to identify something about their appearance that they like, or something they think they can do well. Keep notes as before, then again ask the group who else would like to . . . have curly hair, be able to swim without armbands, build brilliant Lego® models.

- Level I
- Level II
- **Level III**
- Level IV
- **Circle Time**
- Hall/PE
- Literacy
- Topic Work
- Drama
- **Small Group**

© C Delamain &
J Spring 2000.
Photocopiable

DEVELOPING BASELINE COMMUNICATION SKILLS

PERSONAL AND SOCIAL DEVELOPMENT

CONFIDENCE AND INDEPENDENCE

Messengers

Aim
To be able to use the school building and approach less familiar teachers confidently.

An assortment of pictures, one for each child.
Pencil and paper for you.

For this game the cooperation of other members of the school staff is needed. The pictures should be distributed to other staff members at the start of the school day, and the staff warned that some reception children will be sent to collect them. You need to keep a list of which staff member has which picture. The children are sent off individually and told whom they are to find, and what picture they are to ask for. For children who are still extremely shy, or who have problems with speech or language, a written message may be taken.

The children can be given a question to ask the staff member, and they must bring back the reply.

PERSONAL AND SOCIAL DEVELOPMENT

Job to Job

Aim
To be able to move on from one task to another, without prompting.

Four or five group activities – giant floor puzzle, building a tower of giant bricks, building a train track, sticking pictures onto a collage.

The various activities are set out well apart on the floor. Explain to the children that they are to go in turn to each activity and put in *one piece of the jigsaw, one piece of the train track, one stacking brick, one picture on the collage.* **As soon as they have done one piece of that activity, they must move on to the next activity and add one item to that project. The children can either be sent off individually, in pairs or in threes, but stress that they do not have to move on together – only when they have added their item. This continues until everyone has been to all the activities, or until, for example, the brick tower has collapsed or the train track is complete.**

Level I

Level II

Level III

Level IV

Circle Time

Hall/PE

Literacy

Topic Work

Drama

Small Group

© C Delamain &
J Spring 2000.
Photocopiable

DEVELOPING BASELINE COMMUNICATION SKILLS

PERSONAL AND SOCIAL DEVELOPMENT

Confidence and Independence

Egg Timer

Aim
To be able to persist in an activity for a set time, and change activities without verbal prompting.

Equipment
An egg timer.
Three simple worksheets per child, chosen and differentiated to be well within each child's capacity.

How to Play
The children are seated so that they all have a clear view of the egg timer. Ideally a large timer should be used, or have one per table. Each child is given his three worksheets. Explain that the children are to start on the first worksheet, keeping a wary eye on the egg timer, and that the timer will be the magic signal for moving on to the next worksheet. Turn the timer(s) over when the sand has run through, and start the timing process again. No speaking! The game continues until the egg timer sand has run through three times and the children have worked on each of their worksheets.

PERSONAL AND SOCIAL DEVELOPMENT

CONFIDENCE AND INDEPENDENCE

Relay Race

Aim
To be able to play a part as an individual in a team relay race.

Three or four batons and three or four hoops, depending on the number of teams.

The familiar relay race game. Divide the children into teams. The teams line up one behind the other. One hoop is placed in front of each team on the floor, well away from the front person in the team. The front runner in each team is handed a baton. Explain that the front runner must run as fast as possible to the hoop, pass the hoop over his body, and run back to his team, passing the baton to the next runner, and going to the back of the line. The first team to have all its runners returned is the winner.

Level I

Level II

Level III

Level IV

Circle Time

Hall/PE

Literacy

Topic Work

Drama

Small Group

© C Delamain &
J Spring 2000.
Photocopiable

PERSONAL AND SOCIAL DEVELOPMENT

CONFIDENCE AND INDEPENDENCE

Tower of Babel

Aim
To be able to function as a member of a small group within a wider one.

None.

The children are divided into two groups. Tell each group that they are going to be singing a song, but that the other groups will be singing a different one at the same time. These songs are those that the children have practised and are familiar with. Ideally there should be one adult to start off each team. On a count of three, or the 'conductor's' baton falling, everyone starts to sing. This game disintegrates into chaos and laughter quite quickly! It can be played briefly at the start or end of a singing session.

Have more than two groups.

DEVELOPING BASELINE COMMUNICATION SKILLS

PERSONAL AND SOCIAL DEVELOPMENT

CONFIDENCE AND INDEPENDENCE

Put the Tail on the Donkey

Aim
To be able to wear a blindfold, and be prepared for the group to laugh at your efforts.

Large outline picture of a donkey, minus tail, pinned up on the wall or board.
Separate picture of tail, and some means of attachment.

Explain to the children that they will take turns to be blindfolded, and try to put the donkey's tail on in the right place. It is a great icebreaker if you are prepared to be blindfolded and have a go first. Each child in turn is then blindfolded, guided to the donkey picture, and handed the tail to try to attach it accurately.

The other children can be encouraged to give directions 'Up a bit', 'Down a bit' and so on.

- Level I
- Level II
- Level III
- **Level IV**
- Circle Time
- **Hall/PE**
- Literacy
- Topic Work
- Drama
- **Small Group**

© C Delamain &
J Spring 2000.
Photocopiable

Developing Baseline Communication Skills
PERSONAL AND SOCIAL DEVELOPMENT

Level I

96 / Happy–Sad
97 / If You're Happy …
98 / How Do I Feel?
99 / Thank You/No Thank You
100 / Zig's Day

Level II

101 / Angry–Scared
102 / Good People
103 / Who Let the Cat Out?
104 / All Change!
105 / Blues for the Blues

Level III

106 / In a Dark, Dark Cave
107 / Neighbours
108 / Compliments Bag
109 / Sandcastle Game
110 / Pampering Pets

Level IV

111 / Worry Beads
112 / Prize Draw
113 / Party Plan
114 / Listen to My Voice
115 / Spin-a-Word

PERSONAL AND SOCIAL DEVELOPMENT

FEELINGS AND EMOTIONS

Happy–Sad

Aim
To understand the meanings of happy and sad, and be able to match to the correct facial expression.

Equipment

A set of pictures depicting happy/sad (photocopiable set in Activity Resources pp237–238).
A reversible sign, with 'happy' on one side and 'sad' on the other (photocopiable set in Activity Resources pp239–240).

How to Play

The children sit around a table. Spread the pictures out, face down, on the table. Select 2 children (A and B). Child A chooses a picture and shows it to B, who holds the sign and turns it to show the appropriate emotion. B then chooses a picture and the next child (C) adjusts the sign. Carry on around the group until everyone has had a turn at choosing a picture and turning the sign.

PERSONAL AND SOCIAL DEVELOPMENT

If You're Happy . . .

Aim
To be able to identify actions which demonstrate happy and sad.

None.

Everyone stands or sits in a circle. The group says or sings the familiar song, 'If you're happy and you know it . . .'. The words are then changed to 'If you're sad and you know it . . .', accompanied by an appropriate action. Continue the activity by alternating the two emotions and adjusting the actions accordingly.

Happy – give a smile, wave your hand, nod your head, clap your hands, turn and smile (turning to your neighbour), laugh out loud (ho ho ho).
Sad – start to cry, wipe your eyes, make a face, hang your head, hide your face (cover face with hands).

Start by alternating happy and sad. Progress on to a mixture, perhaps two happy and one sad, one happy and two sad, and so on.

Level I

Level II

Level III

Level IV

Circle Time

Hall/PE

Literacy

Topic Work

Drama

Small Group

© C Delamain &
J Spring 2000.
Photocopiable

DEVELOPING BASELINE COMMUNICATION SKILLS

PERSONAL AND SOCIAL DEVELOPMENT

FEELINGS AND EMOTIONS

How Do I Feel?

Aim
To be able to think of a situation which makes a person happy or sad.

The happy/sad pictures as used in **Happy–Sad**, *Level I* (Activity Resources, pp237–238).

The children sit round a table. Shuffle the pictures. Each child has a turn at taking a card from the pile and trying to think of a situation that makes them feel like the emotion on the card. At first you may need to supply choices – if child A chooses a 'happy' card, offer two opposing alternatives.

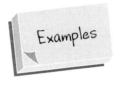

A birthday present *or* a hurt knee
Miss your favourite television programme *or* have burger and chips for tea
Play in the park *or* tidy your bedroom

PERSONAL AND SOCIAL DEVELOPMENT

Thank You/No Thank You

Aim
To be able to match language to simple emotions.

A bag.
An assortment of pleasing toys and useless items, in the bag. There should be more 'nice' items than 'nasty' ones and enough items for each child to take one.

The children sit in a semi-circle around you. Explain that everyone is going to pretend to give each other presents. First child (A) offers the bag to the second child (B) who draws an item from it without looking inside. If it is a 'nice' present, child B must say 'Thank you' with suitable facial expression. If it is a 'nasty' item, child B must say 'No, thank you!', looking suitably disgusted. 'Nice' items are put down beside the child who has drawn them. 'Nasty' items are returned to the bag. At the end, you may like to ask who drew a 'nice' present and who a 'nasty' present, and ask the children to explain why they see them as nice and nasty.

Nice: Doll, toy car, set of colouring pencils, book, hair slide, brooch, items from current popular games.
Nasty: Broken pencil, stone, headless action figure, piece of wood, empty sweet wrapper.

To extend this activity, ask the children to bring in to school and lend some 'nice' and some 'nasty' objects. You will need to explain carefully that they will get their own 'nice' or 'nasty' objects back again!

| Level I |
| Level II |
| Level III |
| Level IV |
| Circle Time |
| Hall/PE |
| Literacy |
| Topic Work |
| Drama |
| Small Group |

© C Delamain & J Spring 2000. Photocopiable

DEVELOPING BASELINE COMMUNICATION SKILLS

PERSONAL AND SOCIAL DEVELOPMENT

FEELINGS AND EMOTIONS

Zig's Day

Aim
To be able to understand what might make another person happy or sad.

Zig (See Awareness of Others, *Level III*, **Alien Visitor 1**). Set of cards which tell what has happened to Zig today. Bag.

The children sit in a semi-circle around you. The cards are put into the bag, and the children take turns to draw one out and hand it to you. Read out what is on the card, and ask the children how they think this has made Zig feel.

Happy
Mummy is icing a cake for his party
Grandma is coming to stay
Tickets for a coach trip arrived
He found a £1 coin
Class won a colouring competition
He got a new bike
His best friend is coming to tea

Sad
His rabbit is not very well
It's wet play again
He's lost his favourite toy
Holiday has been cancelled
Best friend wouldn't sit next to him on the bus
Someone pushed him over

© C Delamain & J Spring 2000. Photocopiable

PERSONAL AND SOCIAL DEVELOPMENT

Angry–Scared

Aim
To understand the meaning of angry and scared, and to be able to match to the correct facial expression.

A set of pictures showing angry/scared (photocopiable set in Activity Resources pp241–242).
A reversible sign showing 'angry' on one side and 'scared' on the other (photocopiable set in Activity Resources pp243–244).

This is played in the same way as Happy–Sad, *Level I.* **Children sit around a table. Spread the pictures out, face down, on the table. Select two children (A and B). Child A chooses a picture and shows it to B, who holds the sign and turns it to show the appropriate emotion. B then chooses a picture and the next child (C) adjusts the sign. Carry on around the group until everyone has had a go at choosing a picture and turning the sign.**

Level I

Level II

Level III

Level IV

Circle Time

Hall/PE

Literacy

Topic Work

Drama

Small Group

© C Delamain &
J Spring 2000.
Photocopiable

PERSONAL AND SOCIAL DEVELOPMENT

Feelings and Emotions

Good People

Aim
To be able to express why we like certain people.

A beanbag.

Divide the group into two teams, who stand in lines opposite each other. In one line (A) the children are each told in a whisper that they have a certain job. The first child in the other line (B) throws the beanbag to a child in line A. The line A child catches the beanbag, and must state what job he has (for instance, a policeman). The children in line B must collaborate to say why they like a policeman. Then the next child in line B throws the beanbag to another line A member, and the process is repeated.
Suggestions for 'jobs':

Policeman	**Doctor**	**Nurse**
Milkman	**Vet**	**Teacher**
Builder	**Librarian**	**Sweet shop keeper**

At a later stage, you may like to introduce some contentious jobs, such as dentist, and encourage discussion as to why the children do or do not like the dentist, and what aspects of the dentist's role are 'good' or 'bad'.

© C Delamain &
J Spring 2000.
Photocopiable

DEVELOPING BASELINE COMMUNICATION SKILLS

PERSONAL AND SOCIAL DEVELOPMENT

Who Let the Cat Out?

Aim
To be able to recognise when to say sorry.

Set of cards which outline 'good' and 'bad' actions. Bag.

The children sit in a semi-circle around you. The cards are put into the bag. Explain that you will be pretending that they have done good or naughty things. If they draw a 'naughty' card, they must say 'sorry'. The children take turns to draw out a card, and you read out what the child is supposed to have done. If it's a 'good' card, praise the child.

Examples

Good
Remembered to feed hamster
Helped mum do the dishes
Played with baby brother
Found dad's glasses
Tidied bedroom

Bad
Let the cat out by mistake
Broke a cup
Hit another child
Left bike outside in the rain
Spilt milk all over kitchen floor

You will know your children and how suggestible they are. If the children's understanding is adequate, the 'bad' scenarios can be made more outrageous and funny (for example 'Brought a muddy elephant into the living room').

Level I

Level II

Level III

Level IV

Circle Time

Hall/PE

Literacy

Topic Work

Drama

Small Group

© C Delamain &
J Spring 2000.
Photocopiable

PERSONAL AND SOCIAL DEVELOPMENT

All Change!

Aim
To be able to recognise that others can have different feelings from oneself.

A list of objects.
Two big cards, one depicting a 'thumbs up' sign and the other a 'thumbs down' (photocopiable pictures in Activity Resources p245).

The 'thumbs up' sign and the 'thumbs down' sign are put on the floor well apart. If this is played in the hall, you might draw two big circles on the floor with the signs in the middle of them. Explain that the 'thumbs up' means they like something, the 'thumbs down' that they do not. You then call something out (see examples in the list below). If they like it, the children go to the 'thumbs up' area; if they dislike it, they go to the 'thumbs down' area. This means that each time you call something out the number of children in either area will probably change.

Medicine	Vegetables
Dolls	Bananas
The colour blue	Any current children's
The colour purple	television programme
Cereal	A current popular toy
Windy weather	Going for walks
Swimming	

© C Delamain &
J Spring 2000.
Photocopiable

DEVELOPING BASELINE COMMUNICATION SKILLS

PERSONAL AND SOCIAL DEVELOPMENT

Blues for the Blues

Aim
To be able to match facial expression and body language to four simple emotions (happy, sad, angry, scared).

Four pieces of coloured card, one blue, one yellow, one red, one black.
Bag or box.

The children sit in a semi-circle in front of you. Put the coloured cards into the bag or box, explaining to the children that each colour stands for a feeling. Blue is for sad, yellow for happy, red for angry, and black for scared. In turn, a child comes to the centre of the circle and takes out a colour card, concealing it from the group. He must then act out the appropriate emotion, while the other children try to guess which emotion it is.

Many of the children will be unable to remember which colour stands for which emotion, and you will have to help by whispering the emotion to the child.

- Level I
- **Level II**
- Level III
- Level IV
- **Circle Time**
- Hall/PE
- Literacy
- Topic Work
- Drama
- **Small Group**

© C Delamain &
J Spring 2000.
Photocopiable

PERSONAL AND SOCIAL DEVELOPMENT

In a Dark, Dark Cave

Aim
To be able to express different feelings about different places.

One giant die.

The group is divided into two teams. Each number on the die is matched to a different emotion as follows: 1= interested, 2 = excited, 3 = amused , 4 = happy, 5 = bored, 6 = scared. A note should be made of this, for reference.

The teams make two lines. The first child from team A rolls the die, and you call out the emotion. The first child from team B tries to think of a place associated with that feeling, and tells the rest of the group. Carry on down the line until everyone has had a go.

If possible write down the responses so that at the end of the game you have a list of interesting, exciting, funny, happy, boring and scary places. This can be a starting point for a class discussion. The children may need help with ideas at first.

© C Delamain &
J Spring 2000.
Photocopiable

PERSONAL AND SOCIAL DEVELOPMENT

Neighbours

Aim
To be able to express and explain own feelings.

None.

Everyone sits in a circle. Explain that each child is going to say 'hello' to the person next to them, and ask them how they are feeling. Choose a child to start (child A). A asks B how he is feeling. B answers, then asks C, and so on round the circle. Encourage children to say *why* they are feeling as they are. At first they will need quite a lot of help to do this; you may need to repeat what they have said, adding 'because …' and then asking them to give the reason.

If child A says 'I'm feeling happy', you should encourage a longer response by replying 'you're feeling happy because …?' Child A might then say '… because, I like school'.

At first they are likely to express very basic emotions, and give very basic answers. As they become more confident, they will be more adventurous, and honest! It is worth doing this activity regularly over a period of half a term or more.

Level I

Level II

Level III

Level IV

Circle Time

Hall/PE

Literacy

Topic Work

Drama

Small Group

© C Delamain &
J Spring 2000.
Photocopiable

PERSONAL AND SOCIAL DEVELOPMENT

FEELINGS AND EMOTIONS

Compliments Bag

Aim
To be able to say something nice about someone else.

A beanbag.

Everyone stands in a circle. Choose a child to start. Child A throws the beanbag to someone (child B). Child B has to say something nice about child A, before throwing the beanbag to child C, who says something nice about B. Carry on until everyone has received a compliment.

- Level I
- Level II
- **Level III**
- Level IV
- **Circle Time**
- **Hall/PE**
- Literacy
- Topic Work
- Drama
- **Small Group**

© C Delamain &
J Spring 2000.
Photocopiable

PERSONAL AND SOCIAL DEVELOPMENT

Sandcastle Game

Aim
To be able to express how it feels to be in a group of people tightly squashed together.

Chalk or lengths of skipping rope.

A game for the hall. Draw two or three big circles on the floor, depending on the number of children in the group, or mark out circles with ropes. Explain that each circle is a sandcastle. As the sea comes in, some sand will get washed away and the circles will get smaller and smaller. The children will have to cluster more and more closely together to stay on the castle. Six or seven children are allocated to each circle. Gradually mark out smaller and smaller circles within the original ones. By the end, some of the children will be unable to fit in to the circle. Each time the circle becomes smaller, ask the children how it feels.

Level I

Level II

Level III

Level IV

Circle Time

Hall/PE

Literacy

Topic Work

Drama

Small Group

© C Delamain &
J Spring 2000.
Photocopiable

PERSONAL AND SOCIAL DEVELOPMENT

Pampering Pets

Aim
To be able to think about the feelings and needs of animals, and talk about them.

Two sets of pictures of pet animals (dog, cat, hamster, guinea pig, goldfish, canary, pony, gerbil, tortoise) as many as you think appropriate.
Two bags or boxes.

The children are divided into two teams, each team seated in a circle. This game requires two adults, one for each team. The first child in each team selects a pet picture. This is passed around from child to child, each one thinking of something the pet needs and likes, and something it would *not* like. When the ideas run out, another pet picture can start on its way round. Your 'team leaders' make a note of the ideas as the game proceeds. When the teams have gone through all the pets, the team leaders compare the ideas.

Likes and Needs	*Dislikes*
Food	Rough handling
Water	Too cold or too hot
Exercise	Being frightened
Quarters kept clean	Wrong sort of food
Brushing/grooming	Nothing to play with

© C Delamain &
J Spring 2000.
Photocopiable

PERSONAL AND SOCIAL DEVELOPMENT

Worry Beads

Aim
To be able to identify things that worry us, and express them.

Box of large threading beads, and lace.

Everyone sits in a circle. The first child is given the lace and selects a bead from the box. He must think of one thing that worries him, and then thread his 'worry bead' on to the lace. He passes the lace and bead box on to the next child. Repeat round the circle.

It is often difficult for children at this age to explain the difference between things that make them sad or frightened, and things that worry them. You may need to start the game off with a worry of your own, to give the children the idea. This game can be expanded later to include other people's worries.

- Level I
- Level II
- Level III
- **Level IV**
- **Circle Time**
- Hall/PE
- Literacy
- Topic Work
- Drama
- **Small Group**

© C Delamain &
J Spring 2000.
Photocopiable

DEVELOPING BASELINE COMMUNICATION SKILLS

PERSONAL AND SOCIAL DEVELOPMENT

Feelings and Emotions

Prize Draw

Aim
To be able to appreciate the humour in ridiculous situations, and use imagination to think how one would feel.

List of 'situations' for adult as a reminder.

The children stand or sit in a circle, adult in the middle. Tell the children they are in a pretend competition and might win a special treat. One child at a time comes to you, and in a whisper you tell him what he has won. The child rejoins the group, and tells them what his prize is, and how he feels about it. Let everyone in the group have a turn.

Go for a flight on a bird's back
Ride on a dolphin
Share a bun with an elephant
Share a banana with a monkey
Turn into a fish for a day
Turn into a cat for a day
Have a magic wand
Have a flying broomstick
Be a clown for a day and tip buckets of water over the other clowns
Be a tightrope walker

© C Delamain &
J Spring 2000.
Photocopiable

PERSONAL AND SOCIAL DEVELOPMENT

Party Plan

Aim
To be able to think of surprising, interesting and exciting things for a party, and discuss the party plans in small groups.

None, except pencil and paper for adult(s).

The class is divided into three groups. If possible, each group has an adult as scribe. One group is to think up as many surprising kinds of food as possible, the next exciting games and the third interesting fancy dresses. The scribes record all the ideas. When ideas run dry, read out the suggestions and lead discussion.

This game makes a good basis for a project.

- Level I
- Level II
- Level III
- **Level IV**
- **Circle Time**
- Hall/PE
- Literacy
- **Topic Work**
- Drama
- **Small Group**

© C Delamain &
J Spring 2000.
Photocopiable

DEVELOPING BASELINE COMMUNICATION SKILLS

PERSONAL AND SOCIAL DEVELOPMENT

Listen to My Voice

Aim
To be able to recognise how someone is feeling from the tone of their voice.

Four different coloured beanbags to represent angry, excited, sad, happy. Use the same colour code as in **Blues for the Blues**, *Level II*, ie, angry = red, sad = blue, happy = yellow, and a new one, excited = purple. A list of words.

The four beanbags are put in different parts of the room. Explain to the children which colour represents which emotion. By now most of the children should be able to associate the colour with the correct emotion, but some will still have to be helped.

Everyone sits or stands in a circle. Choose a child to have the first turn, and say a word in an angry voice. The child goes to the appropriate beanbag. Choose another child and say the same word, but this time in an excited voice. That child goes to the 'excited' beanbag. Continue until all four emotions have been represented, before going on to do the same with the other words. Encourage the children to name the emotion, as well as choosing the right beanbag.

Sausages	Badgers	Thursday
Raining	Treacle	Buns
Everybody	Something	Twenty-two
Green		

© C Delamain & J Spring 2000. Photocopiable

PERSONAL AND SOCIAL DEVELOPMENT

Spin-a-Word

Aim
To be able to join in with you in telling short stories about different emotions.

Short stories (photocopiable sheets in Activity Resources pp246–250).
Spinner which can point to 'boring', 'exciting', 'surprising', 'cross' and 'frightened' (photocopiable sheet in Activity Resources p251).

The children sit in a semi-circle in front of you. The group can use the spinner to choose whether they would like their story to be about 'boring', 'exciting', 'surprising', 'cross' or 'frightened'. When they have chosen, explain that you will be reading a story about that word. Whenever you pause, the children are to shout out the word to fill the gap. (Toby got bored very easily. Whatever his mother or father suggested he might like to do, Toby always answered . . . [boring].)

One or two stories at a time is probably enough. This game can get quite noisy, and is much more fun if it is allowed to be, so it may be necessary to play in the hall!

- Level I
- Level II
- Level III
- **Level IV**
- Circle Time
- Hall/PE
- Literacy
- Topic Work
- Drama
- Small Group

© C Delamain & J Spring 2000. Photocopiable

Developing Baseline Communication Skills
LANGUAGE AND LITERACY

UNDERSTANDING

Level I

118 / If!
119 / Category Bingo
120 / Listen and Colour
121 / Find Zig
122 / What Can it Be?

Level II

123 / Red and Yellow Counters
124 / Zig's Tea
125 / Farmer Fred
126 / Parrot Hunt
127 / Where's Granny Going?

Level III

128 / Musical Messages
129 / Three Clues!
130 / Art Attack
131 / Rats' Tails (i)
132 / Once Upon A Time

Level IV

133 / It's a Funny World
134 / Guess Who?
135 / Listen and Draw
136 / Rat's Tails (ii)
137 / Work it Out!

LANGUAGE AND LITERACY

UNDERSTANDING

If!

Aim
To be able to understand and respond appropriately to a simple classroom instruction.

None.

First make sure you have the whole group's attention. Explain that you are going to ask them some questions. Some of the questions will be really easy, some will be hard. Make sure the children understand that they are only to put their hands up if they DO know the answer. Ask a variety of questions, starting each one with 'If you know … put your hand up'.

Some children will still tend to put their hand up regardless of whether they know the answer or not – this is why it is important to say 'put your hand up' at the end of the question. When hands are up, choose a child to answer the question.

If you know how many legs a spider has, put your hand up.
If you know the first letter of the alphabet, put your hand up.
If you know what colour a banana is, put your hand up.

© C Delamain &
J Spring 2000.
Photocopiable

PAGE 118   DEVELOPING BASELINE COMMUNICATION SKILLS

LANGUAGE AND LITERACY

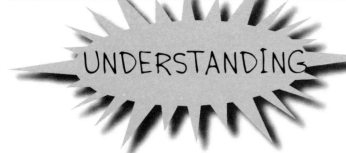

Category Bingo

Aim
To understand six familiar category labels.

A selection of pictures belonging to the six categories: buildings, animals, food, furniture, toys and vehicles – at least two for each category.
Six small cards, each with one category label written on it.

The children sit round a table. Lay the pictures face down on the table. You have the label cards in front of you, face down. Explain that when you call out, they all pick up a card. Choose a label and call out the category, eg, 'animals!' Each child then picks a picture card. If anyone has a picture they think matches the category, they call out 'Bingo!', and if they are right, they keep the card. The remaining cards are replaced face down on the table. Continue until all the cards are used up. The winner is the child with the most cards.

Level I

Level II

Level III

Level IV

Circle Time

Hall/PE

Literacy

Topic Work

Drama

Small Group

© C Delamain & J Spring 2000. Photocopiable

LANGUAGE AND LITERACY

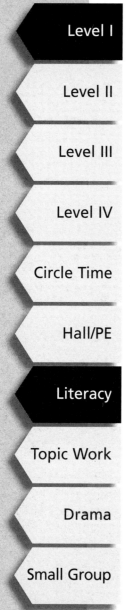

UNDERSTANDING

Listen and Colour

Aim
To be able to understand a simple instruction involving paper, pencils and colouring materials.

Simple line drawing of, for example, a clown – one for each child.
Colouring pencils.

Explain that you want the picture to be coloured in really carefully. Then give instructions to colour one part of the picture at a time, eg, 'Colour the clown's hat blue'. Wait until everyone has finished before giving the next instruction.

Some children will wait and copy others – if possible make a note of this, and if necessary check those children's understanding of the various words in the instruction. They may not understand colour words, or parts of the body or clothes. This activity can be made as simple or difficult as necessary.

© C Delamain &
J Spring 2000.
Photocopiable

LANGUAGE AND LITERACY

UNDERSTANDING

Find Zig

Aim
To be able to understand prepositions relating to space and position.

A doll or creature to represent Zig, the alien visitor. A tiny toy.

Explain to the children that Zig has lost his favourite toy – tell them what it is. Ask them to help Zig look for the toy, by taking him around the class to look in different places. Choose a child to start (child A), and say 'First Zig looked *under* the table'. Child A takes Zig to a table and demonstrates looking under it. Say 'But it wasn't there, so then he looked *in* the cupboard'. Give Zig to child B to look in the cupboard. Carry on like this until every one has had a turn. Then surprise them all by finding the toy in your pocket!

Next time you play the game, find the toy somewhere different. If you have a large group you may have to select only half the group in one session.

Use the following prepositions at first:
in, on, under, in front, behind, beside, next to.

Include harder prepositions: inside, outside, above, below, over, between.

Level I

Level II

Level III

Level IV

Circle Time

Hall/PE

Literacy

Topic Work

Drama

Small Group

© C Delamain & J Spring 2000. Photocopiable

DEVELOPING BASELINE COMMUNICATION SKILLS

LANGUAGE AND LITERACY

What Can it Be?

Aim
To be able to guess an object from a simple definition.

List of simple descriptions of familiar objects.

Explain to the children that you are going to have a quiz. Divide them into two teams, and label them, eg, the red team and the blue team. Use counters or small bricks as 'points' when the team gets the right answer. Describe the first object on your list. Choose a child from the red team to guess what the object is. If the answer is right, the red team gets a point. If not, let the blue team guess.

Orange – it is round, juicy and is a fruit
Elephant – it is an animal, it is big and has a trunk
Tree – it grows, it has branches, you can climb it
Spoon – it is metal, you eat with it, it is shiny
Towel – it is soft, is kept in the bathroom, you dry with it
Light – it has got a switch, it has a bulb, it helps you see
Rabbit – it is furry, it has long ears, you can keep it as a pet
Bike – it has two wheels, you ride it, it is made of metal
Clock – it has a face, it tells you the time, normally it is round
Banana – it is yellow, it is a fruit, it is long

© C Delamain &
J Spring 2000.
Photocopiable

DEVELOPING BASELINE COMMUNICATION SKILLS

LANGUAGE AND LITERACY

Red and Yellow Counters

Aim
To be able to listen and respond to an instruction directed at the group, rather than one individual.

Two sets of different coloured counters or small bricks – enough for everyone.

Give half the group one colour counter, eg, red, and the other half the other colour, eg, yellow. Tell them you are going to give some instructions, but they will need to listen carefully because the people with red counters will get different instructions from the people with yellow counters. The children spread out around the room. Give instructions as follows: 'Go to the door if you have a yellow counter'. Pause while the children carry it out, then give the second instruction 'Sit on the floor if you have a red counter'.

As they become used to the game you will be able to give both groups their instructions in one go – this requires better listening and understanding.

Wave your hand, touch your nose, stand on one leg, touch the table.

Divide the group into three, or even four. This means they have to listen to more, and retain the information that is important for *them*.

Level I

Level II

Level III

Level IV

Circle Time

Hall/PE

Literacy

Topic Work

Drama

Small Group

© C Delamain &
J Spring 2000.
Photocopiable

DEVELOPING BASELINE COMMUNICATION SKILLS   PAGE 123

LANGUAGE AND LITERACY

UNDERSTANDING

Zig's Tea

Aim
To be able to carry out simple instructions to make a drawing.

Paper, pencils.
Toy to represent Zig, the alien visitor

Give each child a piece of paper with a large circle on it, and a pencil. Explain that this is a picture of a plate, and they are going to draw on it what Zig the alien visitor had for tea.

Give the instructions one at a time and wait while the children draw. Collect all the pictures and make sure they are named. At some point later in the day select several at random. Ask the children who produced them if they can remember what they drew.

Zig was very hungry. He had two sausages (pause), a tomato (pause), lots of peas (pause), and a piece of bacon (pause).

© C Delamain &
J Spring 2000.
Photocopiable

LANGUAGE AND LITERACY

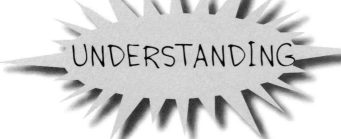

Farmer Fred

Aim
To be able to understand words involving quantity: a few, some, most, lots, all.

A collection of farm animals: cows, pigs, sheep, horses. Five sheets of paper, not smaller than A4 size, with circles of increasing size to represent a few, some, most, lots, all.

This activity works best with a maximum of eight children. If there are more than this, you may need to divide them into sub-groups, each allocated a type of animal to look after. Spread the sheets of paper out on the table or floor, with different numbers of counters in the circles to indicate few, some, most, lots, all. Explain that Farmer Fred needs some help with his animals. Tell them that the sheets of paper are fields, and the area in the middle is the farmyard. Assign a group of animals to each child. In turn, give them instructions as demonstrated below. At first you may need to provide quite a lot of help with this abstract vocabulary. When each child has had a turn, reassemble the animals. Tell the children that Farmer Fred has left all the gates open, and all the animals have escaped into the farmyard! Different children can then be chosen to carry out the next set of instructions.

Put a few pigs in a field
Put lots of cows in a field
Some of the horses need to go in a field
Put all the sheep in a field

Level I

Level II

Level III

Level IV

Circle Time

Hall/PE

Literacy

Topic Work

Drama

Small Group

© C Delamain & J Spring 2000.
Photocopiable

DEVELOPING BASELINE COMMUNICATION SKILLS

LANGUAGE AND LITERACY

UNDERSTANDING

Parrot Hunt

Aim
To be able to listen for specific information while listening to a story.

Equipment

A short piece of text, approximately 200 words, containing specific information to listen for, such as an animal, a food, a toy.

How to Play

The children sit in a semi-circle facing you. Explain that you are going to tell them a little story, and that they need to listen carefully. Tell them that if they listen really carefully they will find out which animal (or food, etc) is in the story. They must not shout out when they hear which animal it is, but wait until you finish the story, and then put their hand up if they know.

Example

There was once a boy called Sam, and he lived with his mum and dad and little sister. One day Sam was playing in the garden when he heard a strange noise. It was coming from the garden shed. He tried to open the shed door, but it was locked. He put his ear against the door and listened. There it was again! It sounded a bit like a whistle, and a bit like a squeak, and then it sounded like someone coughing. What could it be? Suddenly Sam jumped – a croaky voice said 'What do you think *you're* doing?'

Sam walked round to the side of the shed. The window was open a little bit, so he got a stick and managed to force it wide open. Just then his mum called him in for tea. He was really hungry, so he turned round and ran towards the house. After tea he went back to the shed and climbed on a box so he could look through the window. The shed was quiet now, and empty except for the flower pots, the spade and the wheelbarrow. Sam didn't see the green and blue parrot, perched in a tree in the corner of the garden.

© C Delamain &
J Spring 2000.
Photocopiable

LANGUAGE AND LITERACY

UNDERSTANDING

Where's Granny Going?

Aim
To be able to draw inferences.

Lists of things Granny might have with her, as reminders (suggested lists in Activity Resources, p252).

The children sit in a semi-circle in front of you. Explain that Granny has a lot of places to go to. As she collects what she needs for each outing, the children have got to guess where she is going. As soon as someone thinks they know, they must put their hand up, and they can make their guess. If they are right, Granny must change her mind and get ready to go somewhere else. If they are wrong, the list continues until Granny's destination is guessed correctly.

Granny is collecting a towel, some sun tan lotion, a sun hat, a swimsuit.
Granny is collecting a space helmet, moon boots, a space suit.
Granny is collecting her purse, a big bag, a shopping list and her umbrella.
Granny is collecting her tickets, an airline bag, travel pills.
Granny is collecting her warm trousers, a warm jacket, her skates.

This game can also include **'What's the Weather Like?'** (Granny is getting her woolly hat, a scarf, fur boots and a thick jacket. Granny is getting her umbrella, her wellies and a mackintosh.)

- Level I
- Level II
- Level III
- Level IV
- Circle Time
- Hall/PE
- Literacy
- Topic Work
- Drama
- Small Group

© C Delamain & J Spring 2000. Photocopiable

DEVELOPING BASELINE COMMUNICATION SKILLS

LANGUAGE AND LITERACY

UNDERSTANDING

Musical Messages

Aim
To be able to follow more complex spoken instructions.

Instructions written on folded slips of paper.
A tape recorder and music cassette.
A container for the instructions.

Everyone sits in a circle. Explain to the group that you are going to pass the container round while the music is playing. When the music stops, the person with the container takes a 'message' out and passes it you. Read the message and that child responds. Continue until everyone has had a turn.

Tell me something you write (eat, colour, play etc) with
Show me something made of metal (glass, plastic, wood, paper, etc)
Name something that is alive
Think of an animal that can be a pet
Think of a fruit that is yellow and long
Tell me something that is furry and squeaks
Show me something tall
Think of something that feels rough (smooth, silky, soft, hard, etc)
Tell me something that tastes sweet (sour, salty, etc)
Find something that can hold liquid

© C Delamain &
J Spring 2000.
Photocopiable

LANGUAGE AND LITERACY

UNDERSTANDING

Three Clues!

Aim
To be able to work cooperatively to identify an object.

A selection of objects, hidden from sight.
A bag or box.

The children sit in a semi-circle round you. Explain that you are going to play a thinking game. Show them the bag, and say that you are going to hide something in it, then you are going to give them three clues. Choose three children (A, B and C) to remember a clue each.

Put one of the items in the bag, and in a whisper tell A the first clue, B the second, C the third. Encourage A, B and C to tell their clue to the rest of the group. If anyone knows the object they must put their hand up. When the object has been guessed, put another in the bag and choose another three children to remember the clues.

The clues should include what it is made of, looks like, is used for, where you might find it etc.

Objects:
book, spoon, scissors, brush, coffee jar, apple, mirror, key, stone, leaf

Clues:
1 It is made of paper
2 It has pages
3 You can read it

- Level I
- Level II
- **Level III**
- Level IV
- **Circle Time**
- Hall/PE
- **Literacy**
- Topic Work
- Drama
- Small Group

© C Delamain &
J Spring 2000.
Photocopiable

DEVELOPING BASELINE COMMUNICATION SKILLS   PAGE 129

LANGUAGE AND LITERACY

Art Attack

Aim
To be able to carry out instructions involving paper and pencils.

Sheets of paper, divided into six or eight boxes, numbered.
Colouring pencils.
A list of suitable instructions.
A 'master copy' of the finished sheet.

Give each child a sheet of paper and the relevant colouring pencils. Ask them to find box number one. Explain that you are going to ask them to draw something in that box and that they need to listen carefully. Tell them you will say it twice.

Make sure each instruction is restricted to two key items: shape and colour. Continue to give the instructions until all the boxes are filled. Then show them your master copy, and see how many can match theirs to yours.

If they are really good at this you can:
- increase length of instruction, eg, draw a red square and a blue circle
- add another item of information: size, eg, draw a big yellow circle

Draw a red square
Draw a green triangle
Draw a yellow circle

© C Delamain &
J Spring 2000.
Photocopiable

LANGUAGE AND LITERACY

Rats' Tails (i)

Aim
To be able to understand instructions involving long/short.

Ribbons or strings, in two lengths, enough for half the group.

Divide the children into two groups. One group are the 'rats', the other group are the 'catchers'. Give half the rats a long tail, the other half a short tail. The 'rats' tuck one end of their tail into the back of their shorts, trousers or skirts. Explain that the rat catchers are coming, and they might be looking for long tails or they might be looking for short tails.

Then tell the rat catchers you want them to catch rats with either 'short tails' or 'long tails'. When you say 'GO!', the rat catchers chase the rats and try to catch as many of the right kind of tails as possible in a given time, eg, 15 seconds. When you say 'Stop!' all the rats with no tails sit out. Repeat the game, varying the type of tail to be caught, until all the rats have been caught. Swap the rats and catchers, so each group has a chance to be both.

- Level I
- Level II
- **Level III**
- Level IV
- Circle Time
- **Hall/PE**
- Literacy
- Topic Work
- Drama
- Small Group

© C Delamain & J Spring 2000. Photocopiable

LANGUAGE AND LITERACY

- Level I
- Level II
- **Level III**
- Level IV
- Circle Time
- Hall/PE
- **Literacy**
- Topic Work
- Drama
- Small Group

Once Upon a Time

Aim
To be able to listen to a story and retain specific information.

Short stories, lasting no more than two or three minutes.

This is the same sort of activity as Parrot Hunt **(see Understanding** Level II**)**
However, this time the questions can be a little bit more difficult, for instance:
'Who was in Sam's family?', 'What sounds did he hear in the shed?', 'How did he get the window open?', 'What meal did he have?'

There was once a boy called Sam, and he lived with his mum and dad and little sister. One day Sam was playing in the garden when he heard a strange noise. It was coming from the garden shed. He tried to open the shed door, but it was locked. He put his ear against the door and listened. There it was again! It sounded a bit like a whistle, and a bit like a squeak, and then it sounded like someone coughing. What could it be? Suddenly Sam jumped – a croaky voice said 'What do you think *you're* doing?'

    Sam walked round to the side of the shed. The window was open a little bit, so he got a stick and managed to force it wide open. Just then his mum called him in for tea. He was really hungry, so he turned round and ran towards the house. After tea he went back to the shed and climbed on a box so he could look through the window. The shed was quiet now, and empty except for the flower pots, the spade and the wheelbarrow. Sam didn't see the green and blue parrot, perched in a tree in the corner of the garden.

© C Delamain &
J Spring 2000.
Photocopiable

LANGUAGE AND LITERACY

UNDERSTANDING

It's a Funny World

Aim
To be able to spot absurdities and explain them.

Absurd Stories (available in Activity Resources pp253–255).

The children sit in a semi-circle in front of you. Explain that you are going to read a little story, in which some impossible things will happen. If a child spots an impossible thing, he must put his hand up, and he will be asked to explain *what* was impossible, and *why* it was impossible.

Lee and Daniel set off together down the street. They passed a fish walking the other way (Fish can't walk. They haven't got any legs/feet.) The fish smiled at them and said 'Good Morning!' (Fish can't smile, and can't talk.) The pavements were very crowded, and Lee and Daniel were worried the toyshop would be shut before they got there, so they flew over some houses and into the next street (Boys can't fly. They haven't got wings).

Level I

Level II

Level III

Level IV

Circle Time

Hall/PE

Literacy

Topic Work

Drama

Small Group

© C Delamain & J Spring 2000. Photocopiable

DEVELOPING BASELINE COMMUNICATION SKILLS

LANGUAGE AND LITERACY

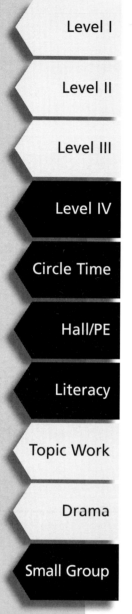

Guess Who?

Aim
To be able to understand spoken information to solve a problem.

None.

The idea is to ask simple questions in order to guess a chosen person, by a process of deduction. Choose two children to be the 'guessers'. They go out of earshot. Select a child from the rest of the group, and tell them who it is. The 'guessers' return and stand in front of the group. They take turns to ask questions relating to hair colour, eye colour, gender and clothing, for example guesser A asks 'Is it a boy?'. If the answer is 'Yes', all the girls sit down. This narrows the search. Guesser B might ask 'Has he got fair hair?', if the answer is 'Yes', all those with brown, black or red hair sit down. Guessing continues until only one child is left standing.

Teacher may need to demonstrate this questioning first, and may need to help some of the children in framing their questions.

© C Delamain &
J Spring 2000.
Photocopiable

LANGUAGE AND LITERACY

Listen and Draw

Aim
To be able to make a drawing that matches a spoken description.

Level I

Level II

Level III

Level IV

 Pencils and paper.
Colouring pencils.

Circle Time

 Give everyone a piece of paper, a pencil and some colouring pencils. Explain that you are going to tell them a little story, and you want them to listen carefully because when you have finished the story you will ask them to draw something.

Hall/PE

Literacy

The stories should be short, descriptive pieces of text, approximately 60 to 70 words in length. Read the story twice, then give the drawing instruction. When they have finished the drawing, ask for volunteers to say what they have drawn. If they have made mistakes, for instance with the colour, find something positive to say about the drawing before bringing their attention to the mistake.

Topic Work

Drama

Small Group

 Josh and his mum went to the fair. There were lots of exciting rides, and Josh bought a toffee apple. He liked the dodgem cars best, and he had three goes. Every time he chose the same car. It was yellow with red wheels. He liked chasing the purple car with white spots on it, and he nearly always hit it!

'Draw a picture of Josh in his car.'

 Make sure that you do not expect the children to remember more than three elements to draw:
'it was *yellow*, with *red wheels*'

© C Delamain &
J Spring 2000.
Photocopiable

DEVELOPING BASELINE COMMUNICATION SKILLS

LANGUAGE AND LITERACY

UNDERSTANDING

Rats' Tails (ii)

Aim
To be able to understand instructions relating to size and colour.

Ribbons or strings in two lengths, two different colours.

This is a slightly more complex version of Rats' Tails (i) **Level III. Divide the children into two groups. One group are the 'rats', the other group are the 'catchers'. Distribute the tails. The 'rats' tuck one end of their tail into the back of their shorts, trousers or skirts. Explain that the rat catchers are coming, and they will be looking for long or short tails, either red or blue.**

Then tell the rat catchers who to catch, eg, 'short red tails'. The children then have to run around and collect as many of the right kind of tails as they can in a given time, eg, 15 seconds. Shout 'Stop'! Then change the instruction. Any rat who loses his tail is out, and goes to sit at the side.

Swap the rats and catchers, so each group has a chance to be both.

© C Delamain & J Spring 2000. Photocopiable

LANGUAGE AND LITERACY

UNDERSTANDING

Work it Out!

Aim
To be able to understand implied information.

Short descriptive texts, (masters in Activity Resources, pp256–257).

The children sit in a semi-circle facing you. Tell them you are going to read a short story about someone they do not know. Explain that they need to listen carefully, because you want them to try and guess who the story is about. Read the story once, and ask the children to put their hand up if they think they know who it could be. Choose a child, and if the answer is correct, encourage the child to say why they came to that conclusion. You may need to read the story more than once.

- Level I
- Level II
- Level III
- **Level IV**
- **Circle Time**
- Hall/PE
- **Literacy**
- Topic Work
- Drama
- **Small Group**

© C Delamain &
J Spring 2000.
Photocopiable

Developing Baseline Communication Skills
LANGUAGE AND LITERACY

LISTENING AND ATTENTION

Level I

140 / Listening Walk
141 / Hunt the Sound
142 / Go Game
143 / Guess the Instrument
144 / Musical Bumps

Level II

145 / Where Am I?
146 / Copy Cat
147 / Listening Feet
148 / Mystery Sounds
149 / Mousie-Mousie®

Level III

150 / High or Low?
151 / Threes
152 / Fruit Salad
153 / Oranges and Lemons
154 / Count the Bears

Level IV

155 / Zoo Game
156 / Finders Keepers
157 / You Got it Wrong!
158 / Colouring Rainbows
159 / Bandstand

LANGUAGE AND LITERACY

LISTENING AND ATTENTION

Listening Walk

Aim
To be aware of sounds in the environment.

None

Take the children for a walk around outside the school building. Explain that you are all going to listen for any sounds that may occur, such as aeroplanes, cars, motorbikes, birds, people calling. Tell the children that they are to walk very quietly. They can walk as a group or in a line; when one child thinks he hears something, he must hold his hand up, and when anyone's hand is up the whole group must stop and stand still. Then the child who puts his hand up first, (or one of the children, if several put their hands up at once) is invited to tell what he heard.

If a second adult is available, do the listening walk with two groups, the other adult keeping a list of the sounds identified by her group. At the end compare lists, and the group which has identified the most different sounds is the winner.

© C Delamain & J Spring 2000. Photocopiable

LANGUAGE AND LITERACY

Hunt the Sound

Aim
To be able to locate the direction from which a sound is coming.

Tape recorder or other continuous noise source.
Several large cardboard boxes.

Hide the sound source under one of the boxes while the children close their eyes. One child is chosen to locate the sound source. Children with very poor listening skills may need to be encouraged at first to put their ear close to each box in turn. Continue until all the group have had a turn.

Increase the number of boxes, and make the children stand some way away from them, and point to the source of the sound.

Level I

Level II

Level III

Level IV

Circle Time

Hall/PE

Literacy

Topic Work

Drama

Small Group

© C Delamain &
J Spring 2000.
Photocopiable

DEVELOPING BASELINE COMMUNICATION SKILLS

LANGUAGE AND LITERACY

Go Game

Aim
To be able to carry out a command on hearing a signal.

None.

The children are seated in a circle around you. Explain that you are going to tell them things to do, but that they must only do what you tell them on the command 'Go!'. You issue your instruction, pause for two to three seconds, and then say 'Go!' Some of the children will fail to wait for the signal; others will have forgotten their instructions by the time the signal is given!

Shut your eyes – go!
Open your eyes – go!
Put one hand up – go!
Put the other hand up – go!
Put both hands down – go!
Touch your nose – go!

Increase the length of time between giving the instruction and saying 'Go!' to four, five or six seconds.

You can use a drum, xylophone, triangle or whistle to give the signal instead of using your voice, and you can make the signal quieter and quieter.

© C Delamain &
J Spring 2000.
Photocopiable

LANGUAGE AND LITERACY

Guess the Instrument

Aim
To be able to identify a musical instrument from its sound.

The class box of musical instruments.
Box, box lid or other object which can form a screen or barrier between child and you.

Give the children plenty of opportunity to handle and experiment with the musical instruments before trying this game. Seat a child either on the floor or at a table facing you. Select three pairs of instruments which sound distinctly different from each other, perhaps two drums, two xylophones, and two sets of maracas. Give the child one set, take the second set, and place the screen or barrier between yourself and the child. Explain that you are going to play one of the instruments, and the child is to guess which one it was. He can demonstrate by playing his matching instrument. Raise the screen and let him see whether he was right or wrong. Then other children have their turns with different sets of instruments.

This game can be made progressively harder by including more instruments, and/or choosing instruments that make nearly similar sounds.

Level I

Level II

Level III

Level IV

Circle Time

Hall/PE

Literacy

Topic Work

Drama

Small Group

© C Delamain &
J Spring 2000.
Photocopiable

DEVELOPING BASELINE COMMUNICATION SKILLS

LANGUAGE AND LITERACY

Listening and Attention

Musical Bumps

Aim
To coordinate a physical response with an auditory signal.

Tape recorder and music tape, piano or guitar.

The familiar party game. The children spread out around the room. Explain that you are going to play some music and while the music is playing, everyone must dance or move around. When the music stops, the children are to sit down on the floor. The slowest person each time will be 'out'. The last one in is the winner.

Some children will never have played this kind of competitive party game. It may be necessary for you to hold the hand of a very uncertain child and carry out the actions with him until he has got the idea and gained confidence.

Musical chairs is a logical extension to this game, and embodies the same listening principle while being a much harder game to grasp.

© C Delamain & J Spring 2000. Photocopiable

LANGUAGE AND LITERACY

Where Am I?

Aim
To be able to locate the source of quiet and/or intermittent sounds.

Sound makers which can play without needing mains leads (battery-operated radio or tape recorder, wind-up or battery-operated musical toys).

Explain to the children that you are all going to play a hide-and-seek game. Tell them what they will be listening for. They must be very quiet and listen carefully. Whoever finds the hidden sound maker must stand still and put his hand up. Tell the children to cover their eyes, and hide the sound maker somewhere around the room. The children open their eyes, and walk or tiptoe quietly round listening for the sound. After the sound maker has been found, repeat using a different hiding place.

Start with a continuous sound at moderate volume. Once the children have learned to observe the rules of the game, and are beginning to use their ears rather than their eyes, you can progressively reduce the volume.

You can make this a good deal harder by using a toy that makes an intermittent sound.

Level I

Level II

Level III

Level IV

Circle Time

Hall/PE

Literacy

Topic Work

Drama

Small Group

© C Delamain &
J Spring 2000.
Photocopiable

DEVELOPING BASELINE COMMUNICATION SKILLS   PAGE 145

LANGUAGE AND LITERACY

LISTENING AND ATTENTION

Copy Cat

Aim
To be able to imitate a sequence of two sounds.

Equipment
Class box of musical instruments.
Screen or barrier (see **Guess the Instrument** *Level I*).

How to Play
Seat a child either on the floor or at a table facing you. Select three or four pairs of instruments. Give the child one set, you keep the other of the pair. Place the screen or barrier between you so that the child cannot see your instruments. Explain that you are going to play two sounds, and the child is to try to match the sequence on his instruments. When the child has played his sequence, raise the screen and show him whether he is right or wrong.

Tip
Use sounds which are quite different from each other at first, such as drum and triangle. As the children become more proficient, the sounds can become harder to distinguish.

LANGUAGE AND LITERACY

LISTENING AND ATTENTION

Listening Feet

Aim
To be able to distinguish between loud and soft, and understand the relevant vocabulary.

Musical instruments.

Before playing this game you will need to have had some discussion about loud and soft sounds, and given the children the opportunity to try out making loud and soft sounds on musical instruments. Divide the children into two groups, each identified by a leader or a colour. Explain that the children are going to 'listen' and 'talk' with their feet. When their feet 'talk' loudly they will be pretending to be something loud and noisy. When their feet talk softly, they will be pretending to be quiet things. Choosing one group (A), play an instrument for them either loudly or softly. The children must either march up and down on the spot as loudly as they can, or tiptoe very quietly on the spot, to match the volume of the instrument. Group B children are asked, for example, 'Were they being elephants or butterflies?' When they have guessed correctly, it is group B's turn to use their listening feet.

Elephants or butterflies   Spiders or big dogs
Hippopotamuses or birds   Lorries or bicycles
Mice or bears   Snow or thunder

- Level I
- **Level II**
- Level III
- Level IV
- **Circle Time**
- **Hall/PE**
- Literacy
- Topic Work
- Drama
- **Small Group**

© C Delamain &
J Spring 2000.
Photocopiable

DEVELOPING BASELINE COMMUNICATION SKILLS

LANGUAGE AND LITERACY

LISTENING AND ATTENTION

Mystery Sounds

Aim
To be able to identify familiar sounds.

A whistle, a cup and spoon, a bunch of keys, a packet of crisps, a tin containing rice or dried peas, one or two musical instruments, two cups of water, pencil, paper, a book.

Let the children investigate the objects and listen to the sounds made by the rustling crisp packet, tearing paper, scribbling on paper, leafing through the book, shaking the tin, pouring water, rattling the keys, stirring the spoon in the cup. Children then sit in a circle round you two by two, with their chairs facing outwards, backs to you. The noisemakers are placed on a table or on the floor in the centre of the circle. One child at a time comes and chooses which noise to make. His 'pair' tries to guess what the sound is. Continue round the group.

Any reasonably recognisable sound can be included. A selection of tins can be used containing dried peas, rice, sugar, one sweet.

LANGUAGE AND LITERACY

LISTENING AND ATTENTION

Mousie-Mousie®

Aim
To detect a specific sound amongst other sounds, and react quickly.

Commercial game Mousie-Mousie®, or a set of six toy rubber mice with long tails, a plastic cup or mug big enough to cover the mouse body, a mat or piece of cloth.

This game is suitable for a maximum of six children. The children are in a small group round the table. Each child is given a mouse. The first child to have a turn places his mouse on the mat or cloth, and takes hold of the end of its tail. Explain that you are the cat, and when he hears you give a 'miaow' you are going to try to catch him by popping the plastic cup down over his mouse. His job is to pull his mouse out of the way the moment he hears the 'miaow'. You will be making lots of other animal noises, but he must wait until he hears the 'miaow' and only then pull his mouse clear. Continue until everyone has had a turn at being the mouse.

This is an extremely useful game for working with letter sounds. Choose a sound that is being worked on in literacy. Tell the children what sound they are listening for (eg, 'ssssss'). Explain that you will make lots of other letter sounds, but today 'sssss' is the dangerous sound and they must jump their mice out of the way when they hear that sound.

- Level I
- **Level II**
- Level III
- Level IV
- Circle Time
- Hall/PE
- **Literacy**
- Topic Work
- Drama
- **Small Group**

© C Delamain & J Spring 2000. Photocopiable

LANGUAGE AND LITERACY

Listening and Attention

High or Low?

Aim
To be able to distinguish between high and low sounds and understand the relevant vocabulary.

Piano, guitar, xylophone or recorder.

Start by playing the children widely contrasted high and low sounds on the instrument or instruments available. Use the terms 'high' and 'low', 'same' and 'different', and encourage the children to try and sing the notes. Children are then separated into two groups. One group is the 'high' group, the other the 'low' group. Explain that when you play a high note, the 'high' group are to stand on tiptoe and reach their arms up into the air. When you play a low note, the 'low' group are to crouch down near the floor. Start playing high and low sounds, clearly differentiated but in random order, until the groups are confident. Then start to play high and low sounds alternately, gradually speeding up, until the two groups are carrying out their actions in a steady alternating rhythm. Continue until the pace becomes too fast. Change over groups.

This activity can be developed by playing notes that are much nearer to each other on the scale, and asking the children to judge whether they were the same or different. A further step still is to play notes very close to each other on the scale, and ask the children to judge whether the second note was higher or lower than the first.

© C Delamain &
J Spring 2000.
Photocopiable

LANGUAGE AND LITERACY

Threes

Aim
To be able to imitate a sequence of three sounds.

Class box of musical instruments.
Screen or barrier (see **Guess the Instrument**, *Level I*).

Seat a child either on the floor or at a table facing you. Select five or six pairs of instruments. Give the child one set, you keep the paired set. Place the screen or barrier between you so that the child cannot see your instruments. Explain that you are going to play three sounds, and the child is to try and match the sequence on his instruments. When the child has played his sequence, raise the screen and show him whether he was right or wrong.

Level I

Level II

Level III

Level IV

Circle Time

Hall/PE

Literacy

Topic Work

Drama

Small Group

© C Delamain &
J Spring 2000.
Photocopiable

LANGUAGE AND LITERACY

LISTENING AND ATTENTION

Fruit Salad

Aim
To be able to listen to your identity as one of a group, and remember it.

None.

The children are divided into small groups of four or five, each group standing well apart, in lines one behind the other. Each group is allocated a fruit – a group of apples, a group of pears, a group of bananas, a group of pineapples, a group of grapes. You stand in front of the groups. Explain that you are going to make some bowls of fruit salad, and you will need some fruit from each group. Call out for a pear (the front child from the 'pear' line should come to you), and an apple, and so on, in random order until you have one 'bowl' of all the fruits. Move a little way away, and call for the fruits to make a second bowl. Continue until all the 'fruit' has been used up. It makes this game more fun if from time to time you call out something thoroughly unsuitable to put in a fruit salad (a frog, a piece of wood). Are the children paying enough attention to spot it?

© C Delamain &
J Spring 2000.
Photocopiable

LANGUAGE AND LITERACY

Oranges and Lemons

Aim
To be able to maintain attention to changing instructions for several minutes and while busy.

Paper and crayons, enough for each child to have several sheets of paper and a shared collection of crayons.

The children are seated in groups at small tables. Explain that some groups are 'oranges' and some are 'lemons', and tell each group their identity. They are to listen carefully to what you are going to tell them to do. Oranges will be doing one thing, lemons something different. After two minutes, you will be giving them all a different job. Give instructions in this form 'Oranges, will you draw a ball and colour it. Lemons, will you draw a house. Off you go.' After one to two minutes, say 'Listen! Now I want the oranges to ...' and so on. The objects you choose need to be quick and easy to draw. This activity should not last more than eight to 10 minutes.

A coloured ball
A house
A face
A star
A square
A tree

Instead of having a whole table as oranges or lemons, which allows some children to copy others, allocate the fruits individually to each child. There are then two or three oranges and two or three lemons on each table.

Level I

Level II

Level III

Level IV

Circle Time

Hall/PE

Literacy

Topic Work

Drama

Small Group

© C Delamain &
J Spring 2000.
Photocopiable

DEVELOPING BASELINE COMMUNICATION SKILLS   PAGE 153

LANGUAGE AND LITERACY

Count the Bears

Aim
To be able to listen for and respond to specific words.

Counters and a small container for each child.

Seat the children facing alternately inwards and outwards in their circle or group, to reduce the likelihood of copying. Explain that you are going to say a list of animal names. Every time they hear the word 'bear' they are to put a counter in their pot or container. You call out the animal names one at a time, including five bears, 'Dog, cat, bear, hippopotamus, bear, elephant, cow, mouse, frog, bear, bird, bear, snake, guinea pig, camel, bear, giraffe.' Tell the children they should each have five counters in their pot. Five counters equals good listening! Repeat with a new list, this time containing a different number of bears, say three or four. Check again – who has the right number of counters?

Speed it up.
Instead of a simple name list, put the animal names into very short phrases 'A black dog, a flying bird, a cow in a field, a bear in the zoo, a frog hopping, a teddy bear'.

© C Delamain &
J Spring 2000.
Photocopiable

DEVELOPING BASELINE COMMUNICATION SKILLS

LANGUAGE AND LITERACY

Zoo Game

Aim
To be able to listen to cues and maintain attention to them for five minutes.

None. The story should be improvised if possible, so that it can be adapted to the observed needs of the group as the game progresses.

The children sit on chairs or mats in a circle round you, with enough room *outside* the circle for the children to run round. Explain that you are going to tell a story about the class going on a visit to the zoo. When a child hears his own name mentioned in the story, he must get up and run once around the outside of the circle. When the word 'zoo' is heard, the whole class must get up and run round, *in the same direction*. This game is best played with a maximum of 10, so that every child can have several turns. It allows you to spot the child whose attention is wandering, and bring his name into the story immediately. As the story progresses, more than one child can be named at a time.

Mrs Smith's class was going on a coach trip to the *zoo*. [Everyone should get up and run round the circle.] The coach driver sorted out who was going to sit at the back. *Melanie* got on first. [Melanie gets up and runs round.] *Katie* wanted to sit next to her, so she got on next. *Darren* wanted to be by a window – and so did *Kevin*. When they were all sitting down with their seat belts done up, the coach set off for the *zoo*. *Sally* had brought some sweets, so she handed them round. *John* and *David* had learnt a new song, so they taught it to the others. It was a long drive, and *Daniel*, *George* and *Alice* got bored. But at eleven o'clock the coach arrived at the *zoo*. And so on, until the concentration span of the group is exhausted!

Level I

Level II

Level III

Level IV

Circle Time

Hall/PE

Literacy

Topic Work

Drama

Small Group

© C Delamain &
J Spring 2000.
Photocopiable

DEVELOPING BASELINE COMMUNICATION SKILLS

LANGUAGE AND LITERACY

LISTENING AND ATTENTION

Finders Keepers

Aim
To be able to listen to an instruction, and wait to carry it out until told.

Pencil and paper for you, to note down what each child has been told to find.

The children sit in a circle. Explain that each child will be told to go and find a certain item in the room. They must wait until everybody has been told what to find before they move. You will say 'Now!' to tell them when to start. When they have found their object, they must return to the circle and sit down. Each child is named in turn and told to find an everyday object ('Christopher, will you find a red crayon. Susie, will you find a pot of glue'). When the whole group has been given an object to find, say 'Now!' and the hunt begins. As the objects are brought, they can be put in a pile on the floor in the middle of the circle.

This game is best played with not more than eight or 10. Children known to have short concentration spans or poor memories should be among the last to be given their instructions.

© C Delamain & J Spring 2000. Photocopiable

LANGUAGE AND LITERACY

You Got it Wrong!

Aim
To be able to detect deliberate mistakes in a story.

 Any short storybook with which the children are familiar.

 **The children are seated in a semi-circle in front of you. Explain that you are going to read them a story (tell them its name) but you think you may be going to make some mistakes. Can they help you put it right? If they spot a mistake, they should shout 'Stop! You got it wrong!' and you will choose someone to tell you what was wrong.**

 One day, Little *Blue* Riding Hood was sent by her mother to take some goodies to her *grandfather*. Little Red Riding *Hat* set off through the woods. It was a lovely day, and she dawdled along the way, and stopped to pick some *poisonous toadstools* which she put in her basket. Little Red Riding Hood didn't know that following behind her, hiding behind the trees, was the wicked *tiger* . . .

Level I

Level II

Level III

Level IV

Circle Time

Hall/PE

Literacy

Topic Work

Drama

Small Group

© C Delamain &
J Spring 2000.
Photocopiable

LANGUAGE AND LITERACY

Colouring Rainbows

Aim
To be able to stop and start working on command in a shared project.

Pictures of rainbows to colour in, one for each child. They should have seven bands for the seven rainbow colours, and a coloured dot by each band to indicate the right colour for that band.
Crayons.

The children are seated at their tables, each child with a rainbow picture, and a shared supply of crayons. Explain that they are going to help each other colour the rainbows in. When you say 'Start!' they should begin to colour their rainbow. When you say 'Stop and pass!' they should stop colouring, and pass their picture on to their right-hand neighbour. Repeat until all the children have contributed to all the rainbows at their table.

It may be helpful to use a musical signal for 'Stop and pass!' such as striking a triangle or blowing a whistle.

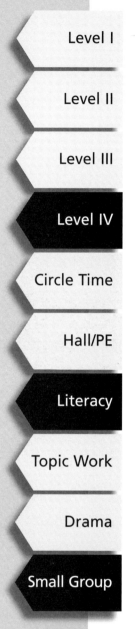

© C Delamain &
J Spring 2000.
Photocopiable

LANGUAGE AND LITERACY

Bandstand

Aim
To be able to imitate a sound sequence as part of a group.

Musical instruments.

Group the children in threes. In group A each child has a triangle, in group B each child has a drum, and so on until each group has been allocated a different instrument. Explain that you are going to play a 'tune' consisting of a sequence of three different instruments. The children must listen, and then copy the sequence group by group. Repeat, varying the sequence each time, and after a few turns swap the instruments around or introduce new ones.

You play a sequence consisting of a drum beat, then a shaken maraca, and then a clash of cymbals. The group with the drums should go first with a drum beat, then the maraca group should shake their maracas, and lastly the cymbals group should give a clash on their cymbals.

This is quite likely to lead to some heated discussion, which can be directed by you to help the children work out what went wrong and why.

Stand the children with their backs to you so they cannot use vision.

- Level I
- Level II
- Level III
- **Level IV**
- **Circle Time**
- **Hall/PE**
- Literacy
- Topic Work
- Drama
- Small Group

© C Delamain & J Spring 2000. Photocopiable

Developing Baseline Communication Skills
LANGUAGE AND LITERACY

SPEAKING

Level I

162 / Something's Missing
163 / Big Green Apples
164 / Circus Act
165 / Do As I Say
166 / Raindrops

Level II

167 / Zig's Day
168 / Colour Families
169 / George the Giant
170 / Zig's Picnic
171 / Picture Partners

Level III

172 / Obstacle Course
173 / Our Own Story
174 / Spot the Difference
175 / Disguises
176 / Only One Left

Level IV

177 / Zig at the Zoo
178 / Make it Up
179 / Gold Crowns
180 / Oops!
181 / Imagine It

LANGUAGE AND LITERACY

Something's Missing

Aim
To name objects confidently in a group.
To be able to ask each other 'What?' questions, and reply.

Equipment

10 everyday objects.
A cloth large enough to cover them when they are laid out on the table.

How to Play

This is basically Kim's Game, so has a secondary value in training visual memory. Spread the objects out on the table, and cover them with the cloth. Tell the children that you will want them to look very carefully at the objects, so that they can remember them. Lift the cloth and start timing (give them up to 30 seconds). Then cover the objects with the cloth again. How many can the children remember? Let them answer in any order or all together. You keep count on your fingers until all the items have been mentioned.

For the second half of the game, cover the objects up with the cloth again. Tell the group to close their eyes. Feel under the cloth and remove one object, which is put out of sight. Then tell the group to open their eyes, and remove the cloth, asking 'What's gone?' or 'What's missing?' The children try to guess. Once the rules of the game are established, the children can take turns to act as 'teacher'.

© C Delamain &
J Spring 2000.
Photocopiable

LANGUAGE AND LITERACY

Big Green Apples

Aim
To be able to describe a picture using basic adjectives.

Two identical sets of pictures. Each set contains toys, fruits and flowers. Each toy, fruit or flower has four different representations (big green apple, big red apple, little green apple, little red apple) (photocopiable masters to be coloured in Activity Resources p258). This game needs two adults, one for each group.

Divide the children into two groups, who sit a little apart. Give each group one set of pictures. The groups should not be able to see each others' pictures. Each group chooses a child to be their leader. Group A selects a picture and their leader holds it in his hand so that the other group cannot see it. He tells group B what it is. 'It's a big green apple' (or 'a little red flower'). Group B try to find their matching picture, and their leader holds it up. Does it match? Then swap groups, group B choosing the picture and describing it, group A trying to find the match.

Level I

Level II

Level III

Level IV

Circle Time

Hall/PE

Literacy

Topic Work

Drama

Small Group

© C Delamain & J Spring 2000. Photocopiable

LANGUAGE AND LITERACY

Circus Act

Aim
To be able to produce short sentences including a range of action words.

A finger puppet for each child, representing clowns.

Give out the finger puppets, including one for you. Explain that everybody is in charge of a clown in a circus. You start the activity by saying, for example, 'My clown can jump through a hoop'. Turn to the first child and say 'What can your clown do?' The child responds, and asks the next child, and so on until everybody has had a turn.

At first the responses are likely to be simple, eg, 'My clown can jump' and some children may tend simply to copy others' ideas. If this persists, you should take another turn and introduce some more variety, eg, '*My clown can balance a bucket of water on his head*'.

LANGUAGE AND LITERACY

Do As I Say

Aim
To be able to tell each other to carry out simple actions.

Set of action pictures, in a bag.

Children stand in a circle. Explain that they are going to take turns to tell each other what to do. You will give the 'caller' a picture showing what the action is to be. The first child is chosen as 'caller'. Hold out the bag for him to withdraw a picture. He must then tell the group what to do. He should be discouraged from demonstrating! Give each child one or more turns to be 'caller'.

Sort out the pictures beforehand, so that you only include actions that can be carried out on the spot and which do not need 'props'. Pictures of somebody running, or riding a bicycle, are not practical!
There are several commercial sets of action pictures available. The ColorCards® series is in widespread use in schools.

- Level I
- Level II
- Level III
- Level IV
- Circle Time
- Hall/PE
- Literacy
- Topic Work
- Drama
- Small Group

© C Delamain &
J Spring 2000.
Photocopiable

DEVELOPING BASELINE COMMUNICATION SKILLS

LANGUAGE AND LITERACY

Raindrops

Aim
To be able to ask each other 'Where?' questions and reply.

A large simple drawing of a house with a chimney, upstairs and downstairs windows, a door and garage, a tree, a fence and a gate. Teachers with artistic talent may like to add a cloud, a pond, a car, a bird.
A box of small coloured round stickers.

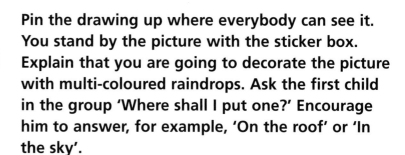

Pin the drawing up where everybody can see it. You stand by the picture with the sticker box. Explain that you are going to decorate the picture with multi-coloured raindrops. Ask the first child in the group 'Where shall I put one?' Encourage him to answer, for example, 'On the roof' or 'In the sky'.

Apply a sticker as instructed, and move on to the next child. Once the rules of the game are established, choose children in turn to be the 'teacher'. The game can continue until the picture is covered with a heavy shower of raindrops.

Make sure that the children do not sit near enough the picture to be able to get by with pointing.

© C Delamain &
J Spring 2000.
Photocopiable

LANGUAGE AND LITERACY

Zig's Day

Aim
To be able to recount a simple event to each other.

Zig, the alien visitor toy.
Some doll's clothes to fit Zig (a hat, a scarf, some trousers).
Some personal items (a hairbrush, a toothbrush, a hand mirror, a handkerchief).

You sit at a small table with a chair for Zig. Half the children sit nearby where they can watch what Zig does (group A). The rest of the children sit with their backs turned (group B). You make Zig carry out a simple action, such as falling off his chair, or putting on his hat. One child from group A is chosen to tell group B what happened ('He fell off his chair'). A child from group B is chosen to come and make Zig repeat the action.

This activity is intended to foster natural expression rather than correct grammar, and child-to-child talk rather than adult-to-child talk. If a child says, for instance, 'he falled off his chair', accept it, do not repeat it correctly.

Extend the range of things that Zig does, or give him a sequence of actions ('Blew his nose and stood on his chair').

Level I

Level II

Level III

Level IV

Circle Time

Hall/PE

Literacy

Topic Work

Drama

Small Group

© C Delamain &
J Spring 2000.
Photocopiable

LANGUAGE AND LITERACY

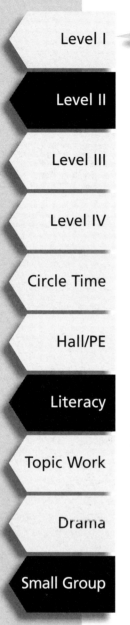

Colour Families

Aim
To be able to ask each other for the colour cards needed, in a simple competitive game with adult support.

Sets of four cards of each colour (four red, four blue, four purple etc.)
Enough colours for all the players to have a different one (templates for colouring available in Activity Resources p259).
A box for each child, marked or coloured to indicate the colour that child should be collecting.

A game for between three and six players. Give each child four assorted cards which they place in their box and conceal from each other. Explain that they have to try and collect four cards of the colour that matches their box. The first child asks his neighbour if he has a card of the colour he wants. If the answer is 'No', that child's turn is over. If the answer is 'Yes', the card is handed over and the player can ask the next child. His turn continues until he receives a 'No'. The winner is the first child to have collected four cards of the same colour.

In the regular Happy Families game, players can ask anyone they like for the card they want. At the level of the game described here, it may be easier just to ask round the circle in turn.

© C Delamain &
J Spring 2000.
Photocopiable

LANGUAGE AND LITERACY

George the Giant

Aim
To be able to ask and respond to a 'Who?' question.

Two identical sets of picture cards (available in Activity Resources pp260–262).
A bag.

Tell the children about a giant called George who was always losing things. Explain that he lived in a castle with a lot of naughty elves, who tricked him by hiding things. Choose one child (A) to be George. The rest of the group are the elves. Give each elf a card, out of George's sight. Put the second set of cards in a bag. George takes a card out of the bag, looks at it, and asks the group, 'Who took my *helicopter*?' The child who has that picture owns up by saying 'I did', and then that child takes the part of George. Continue until everyone has had a turn at playing the part of George.

Remember to remove the two 'used' cards each time.

- Level I
- Level II
- Level III
- Level IV
- Circle Time
- Hall/PE
- Literacy
- Topic Work
- Drama
- Small Group

© C Delamain &
J Spring 2000.
Photocopiable

DEVELOPING BASELINE COMMUNICATION SKILLS

LANGUAGE AND LITERACY

Zig's Picnic

Aim
To be able to describe own actions.

Basket of toy food.
Zig and three or four toys such as a teddy, a doll, a toy animal.

The children sit in a large circle. Seat the toy animals and Zig on the floor in the middle of the circle. Explain that the children are going to give the animals their picnic food. Start the activity by taking the basket of food, choosing an item and saying 'I'm going to give a banana to Zig'. Carry out the action, return to your chair, and give the basket to the first child. The child is then encouraged to give a food item to one of the toys, verbalising his action as modelled by you. He returns to his chair, and passes the basket to the next child. Continue until everyone has had a turn.

Any attempt to produce a sentence should be accepted at this stage, even if there are errors of grammar.

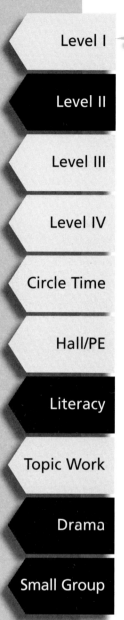

© C Delamain &
J Spring 2000.
Photocopiable

DEVELOPING BASELINE COMMUNICATION SKILLS

LANGUAGE AND LITERACY

Picture Partners

Aim
To be able to explain the relationship between pairs of familiar words.

A set of pictures of 'things that go together', enough to give one to each child in the group.

Sort the cards so that the pairs are separated into two piles. Divide the children into two groups. Children in group A stand in a line side by side. Give them each a card from the first pile of pictures, which they hold up in front of them. Deal the second deck to group B. Tell group B to go and stand beside their 'picture partner' in group A. Some children will need a little help in making the connections. Group B take turns to explain why their picture partners go together.

At the end of the game, each deck may be shuffled and re-dealt, and the teams reversed.

Knife and fork – Because we eat with them
Hammer and nails – You bang nails with a hammer

Accept any attempt at explanation from a child, but if necessary model a more complete answer ('Bang it' – 'Yes, you bang in nails with a hammer').

© C Delamain &
J Spring 2000.
Photocopiable

LANGUAGE AND LITERACY

Obstacle Course

Aim
To be able to give simple instructions involving prepositions.

PE apparatus.

The children are separated into pairs. Explain that one child in each pair is going to be the instructor and one the gymnast. You need to demonstrate this activity with one child first. The instructor is allowed to choose a maximum of three items of apparatus at a time, and tells his partner, the gymnast, what to do with each item. The instructions are given one at a time (see examples below). When the demonstration is complete, the first pair of children take their turn. Encourage the children to think of different sequences and actions. It is important that every child has an opportunity to be both instructor and gymnast, even if it is not possible to manage this in one session.

'Walk along the bench' (wait until the action has been carried out), 'do a somersault on the mat' (wait), 'swing on the rope' (wait).

© C Delamain & J Spring 2000. Photocopiable

DEVELOPING BASELINE COMMUNICATION SKILLS

LANGUAGE AND LITERACY

Our Own Story

Aim
To be able to think up original events in an invented story and relate them.

None.

The children sit in a circle round you. Explain that you are going to make up a story together. You will start it, and then each child will add a little bit on. You might start off something like this 'One day, Paddington Bear decided to go on holiday. He got out his suitcase and . . .'

Some children will need a good deal of prompting with ideas. Others will go on indefinitely, and it is a good idea to have something to hold up as a 'stop' sign when you feel a turn should finish. The children should be encouraged to stick to the story-line as far as they are able.

Level I

Level II

Level III

Level IV

Circle Time

Hall/PE

Literacy

Topic Work

Drama

Small Group

© C Delamain & J Spring 2000.
Photocopiable

LANGUAGE AND LITERACY

Spot the Difference

Aim
To be able to explain similarities and differences.

A set of 'Spot the Difference' pictures enlarged if necessary, or put on an overhead projector (sample pair of pictures available in Activity Resources pp263–264). Two different coloured highlighter pens, one for each team.

Pin or stick the pictures up side by side. Divide the children into two teams. Make sure everyone can see the pictures. Explain that the teams are going to take turns to try and spot a difference between the two pictures. Choose a team to start. Tell them that if anyone can see a difference they are to put their hand up (no calling out). Choose a child whose hand is up to tell you what he has seen. Give help if necessary, but encourage children to try to produce a verbal explanation. If there is general agreement, the child comes up and marks the difference with a dot on one of the pictures. Repeat with the second team. Continue until all the differences have been spotted. Count the coloured dots. The winning team is the one who spotted the most differences.

Spot the difference pictures are available in most good comics and puzzle books.

© C Delamain & J Spring 2000. Photocopiable

LANGUAGE AND LITERACY

Disguises

Aim
To be able to ask and respond to 'Why?' questions.

Enough dressing up items for half the group: dark glasses, hats, scarves, moustaches.

Divide the group into two teams, A and B. Give team A the dressing up clothes, and let them each put on an item. Both teams now line up, facing each other. The first child in team B asks the child opposite him, 'Why are you wearing sunglasses?' The child from team A gives a response, then the next B team child has a turn. Continue until everyone has had a turn. If there is time, swap teams so that team B has a chance to dress up. If this is not possible, reverse the teams in the next session.

You may need to model the responses at first, encouraging the children to try to think of imaginative answers.

Level I

Level II

Level III

Level IV

Circle Time

Hall/PE

Literacy

Topic Work

Drama

Small Group

© C Delamain &
J Spring 2000.
Photocopiable

DEVELOPING BASELINE COMMUNICATION SKILLS

LANGUAGE AND LITERACY

Only One Left

Aim
To be able to ask questions involving adjectives and categories.

Two identical sets of pictures. Each set contains toys, fruits and flowers. Each toy, fruit or flower has four different representations (big green apple, big red apple, little green apple, little red apple) (sets available in Activity Resources p258).
This game needs two adults, one for each group.

Divide the children into two groups. Give each group one set of pictures. The groups should be seated so that they cannot see each others' pictures.

Explain that this is a guessing game. Group A will choose a picture from their own set. Group B must ask questions until they think they can guess which one has been chosen. The adult with group B will probably have to prompt with suitable questions to begin with ('Is it a fruit?' 'Is it a toy?' 'Is it a flower?' 'What colour is it?' 'Is it big or little?') As the questions are answered, with your help group B gradually eliminates one set of pictures after another and pushes them to one side. When they are left with one picture, they choose a speaker to ask 'Is it the little red flower?' If they are right, they win a point. If they have guessed wrongly, the point goes to the other group. Swap over, so that group B now chooses a picture and group A does the guessing.

© C Delamain &
J Spring 2000.
Photocopiable

LANGUAGE AND LITERACY

Zig at the Zoo

Aim
To be able to predict verbally.

Zig, the alien visitor toy
Narrative script (available in the Activity Resources pp265–267).

Explain that Zig often goes to the zoo, so he will be having lots of adventures. The children are going to help you guess what will happen to him. Divide the children into groups of four or five, standing in different parts of the room. Make sure that everyone can both see and hear you. Each group of children pretends to stand by a different animal enclosure. Tell the groups which animals they are standing near. Then read the narrative, pausing where indicated, and asking the appropriate group of children what they think will happen next.

The game can be repeated by changing the groups around, as different children will come up with different ideas.
Alternative narratives can be devised if needed.

- Level I
- Level II
- Level III
- **Level IV**
- Circle Time
- **Hall/PE**
- Literacy
- Topic Work
- **Drama**
- Small Group

© C Delamain &
J Spring 2000.
Photocopiable

LANGUAGE AND LITERACY

SPEAKING

Make it Up

Aim
To be able to construct a simple narrative.

A good selection of pictures of single objects.

Spread the pictures out face upwards on a table. Explain that everyone is going to make up a short story using two of the pictures to give them ideas. Go to the table and choose any two pictures. Then tell a story consisting of two or three sentences involving both pictures. The children then take turns to choose two pictures and tell their own short story.

Picture of sheep and car – 'We saw some sheep when we were in the car'.
Picture of ice cream and puddle – 'I dropped my ice cream in a puddle'.

- Level I
- Level II
- Level III
- Level IV
- Circle Time
- Hall/PE
- Literacy
- Topic Work
- Drama
- Small Group

© C Delamain &
J Spring 2000.
Photocopiable

DEVELOPING BASELINE COMMUNICATION SKILLS

LANGUAGE AND LITERACY

Gold Crowns

Aim
To be able to give key information about an object.

A collection of objects – one for each member of the group.
Counters to represent the 'gold crowns'.

Put the objects out of sight. Tell the children they are going to join in a guessing game, and the idea is to win as many 'gold crowns' as possible. Give each child three crowns before starting. Everyone sits in a circle, or around a table. You model the activity by taking the first turn. Select an object from the collection, but do not let anyone see it. Explain that you will give two clues about the object, and no one must say anything until you have finished. When you have given the clues, anyone who thinks they have the answer puts their hand up. If they are right they gain a 'crown', if not they lose one. Additional clues must be 'bought' with gold crowns. When the first object has been guessed, the child who guessed it takes a turn at giving clues about a different object.

Do not let the children select the objects because they will see the whole collection, and this will make the guessing too easy.

Level I

Level II

Level III

Level IV

Circle Time

Hall/PE

Literacy

Topic Work

Drama

Small Group

© C Delamain &
J Spring 2000.
Photocopiable

LANGUAGE AND LITERACY

SPEAKING

Oops!

Aim
To be able to explain absurdities.

Equipment
Assortment of classroom objects (see examples below).

How to Play
The children sit in a semi-circle facing you. You have a range of classroom objects in front of you. Explain that you were in a hurry that morning, and you did some funny things. Tell them to watch you carefully and see what you did. Then recount your morning's activities one at a time, miming each one using inappropriate props. After each mime, ask the children what you did wrong, and choose a child to explain. Give as many children as possible an opportunity to contribute.

Examples
'This morning I brushed my hair' (mime, using, for example, a ruler instead of a brush).
'Then I phoned the bank' (mime, using, for example, a spoon instead of the telephone).
'Then I buttered my toast' (mime, using a pencil instead of a knife.)

Tip
Encourage the children not only to point out what is wrong, but to suggest what you should be using.

© C Delamain &
J Spring 2000.
Photocopiable

LANGUAGE AND LITERACY

Imagine It

Aim
To be able to put imaginative ideas into words.

A coloured die.
Scenarios written on small cards, enough for one for each child.

This is an activity for a group of not more than 12.

Code the cards with a coloured dot, making sure that you use only colours that appear on the die. Put the cards on the table with the coloured dots visible. Children take turns to roll the die, and select a card of the same colour. Read the card to the child, and ask 'What would it be like…?' Encourage and help the child to share his ideas with the group. This might generate wider discussion, which should be encouraged. Continue until every child has had a turn.

In a submarine
On a mountain top
In an aeroplane
In a cave
On a desert island
On a cloud

Level I

Level II

Level III

Level IV

Circle Time

Hall/PE

Literacy

Topic Work

Drama

Small Group

© C Delamain &
J Spring 2000.
Photocopiable

DEVELOPING BASELINE COMMUNICATION SKILLS

Developing Baseline Communication Skills
LANGUAGE AND LITERACY

Level I

184 / Shopping Game (i)
185 / Farmers and Mechanics
186 / Magic Passwords
187 / Things to Do
188 / Postbag (i)

Level II

189 / Shopping Game (ii)
190 / Matching Pictures (i)
191 / Chinese Whispers (i)
192 / Colour Jumping
193 / Magic Carpets

Level III

194 / Pets' Corner
195 / Whose News?
196 / Chinese Whispers (ii)
197 / Ring me Up!
198 / Postbag (ii)

Level IV

199 / Parrots
200 / Pantomimes
201 / Fact Finder
202 / Matching Pictures (ii)
203 / Jack and Jill

LANGUAGE AND LITERACY

Shopping Game (i)

Aim
To be able to remember two objects for long enough to collect them, and to tell what you have collected.

Assortment of familiar objects (not more than 10). Shopping bag.

Set up a 'shop' with the objects spread out on a table. One adult is the 'shop keeper' and stands by the shop. A second adult sits with the children. The first child is given the shopping bag, and one adult asks 'Please go to the shop and fetch me a . . . and a . . .'. The child is then sent off to collect his shopping. He should bring the objects back in his bag, and without looking inside report what he has collected.

It is crucial that the instruction be given *all at once* before the child is sent off to the 'shop', so that he has to retain the two items in his memory. Encourage the children to repeat their instructions on the way to the shop. Also encourage them to pick up the items in the order they were given.

LANGUAGE AND LITERACY

Farmers and Mechanics

Aim
To be able to remember a two-part instruction.

The classroom collection of toy vehicles and toy animals. Farm buildings and fences, garage, road or railway sets.

Divide the children into two groups of five or six. Tell the children in one group that they are the farmers, and the children in the other group that they are the mechanics. Explain that the farmers are going to sort their animals out, and the mechanics are going to get their cars and lorries tidied up. The farmers have farm buildings, fields, fences, and the toy animals; the mechanics have a garage, road or railway set, and the vehicles. The children take turns. Instruct a farmer 'Put a cow into the field'. Then to a mechanic, 'Put a lorry on to the roundabout'.

If an adult is available for each group, the game can be played as a race. The farmers have ten animals to place, and the mechanics ten vehicles. The winning group is the one who finishes first.

Remember to restrict your prepositions to those with which the children are likely to be familiar (in, on, under, behind, in front, beside or next to). This game can be played using play mats, but this restricts the number of prepositions that can be used.

- Level I
- Level II
- Level III
- Level IV
- Circle Time
- Hall/PE
- Literacy
- Topic Work
- Drama
- Small Group

© C Delamain &
J Spring 2000.
Photocopiable

DEVELOPING BASELINE COMMUNICATION SKILLS

LANGUAGE AND LITERACY

Magic Passwords

Aim
To be able to remember a given word for several minutes.

None.

Five minutes before playtime, lunch time or the end of the school day, announce to the class that you are going to give them a secret magic password. This will allow them to leave the classroom when the time comes. Tell them the word. It can be a real or a nonsense word, but must be something that the children can manage to say. Nonsense words are harder to remember than real ones, so if you make up a nonsense word let the children have some practice saying it. When the time comes for the children to go out to play or to lunch, they must line up and file past you, and whisper the password in your ear. If they get it right, they can 'escape' from the classroom.

Teachers will know which children might become distressed if they cannot remember their password and fear that they might miss playtime or lunch. Suitably reassuring measures should be taken!

© C Delamain &
J Spring 2000.
Photocopiable

LANGUAGE AND LITERACY

Things to Do

Aim
To be able to remember and carry out two linked instructions.

None.

Divide the class into two teams, placed at opposite sides or ends of the hall. This game is played like a relay race. Each member of each team takes his turn and then goes to the back of his line. The first team in which everyone has had a turn is the winner. Two adults will be needed, one for each team. On a count of one, two, three (or ready, steady, go) the first child in each team is given an instruction. The instructions can be different for the two teams.

Go to the wall bars, and touch them
Put your hand up and turn round
Touch your nose and clap your hands
Touch the floor and touch your knees
Walk to the window and jump

Level I

Level II

Level III

Level IV

Circle Time

Hall/PE

Literacy

Topic Work

Drama

Small Group

© C Delamain &
J Spring 2000.
Photocopiable

LANGUAGE AND LITERACY

Postbag (i)

Aim
To be able to remember a series of items and add to it.

An assortment of objects or pictures on the floor in the middle of the circle.
A bag.

This is a 'starter' version of a familiar game, using visual support. The children sit in a circle. Explain that you are going to help the postman remember what he has got in his bag. Start the game off by choosing one of the objects in the collection, putting it in the bag, and saying 'In my bag I've got a . . .'. The bag is passed to the first child, who chooses another object to put in the bag, and says 'In my bag I've got a . . . and a . . .'. The bag is passed round the group, each child repeating the items already chosen, and adding one of his own. The game stops when the list becomes too long for anyone to remember.

It makes this game easier still if you choose objects or pictures which are related to each other in some way, for instance all toys, all pictures of things to do with summer holidays, all clothes.

© C Delamain &
J Spring 2000.
Photocopiable

DEVELOPING BASELINE COMMUNICATION SKILLS

LANGUAGE AND LITERACY

Shopping Game (ii)

Aim
To be able to remember three or four objects for long enough to collect them, and to tell what you have collected.

 Assortment of familiar objects (not more than 12). Shopping bag.

 Set up a 'shop' with the objects spread out on a table. One adult is the 'shopkeeper' and stands by the shop. A second adult sits with the children. The first child is given the shopping bag, and adult asks 'Please go to the shop and fetch me a . . . , a … and a . . .'. The child is then sent off to collect his shopping. He should bring the objects back in his bag, and without looking inside report what it is that he has collected.

 It is crucial that the instruction be given *all at once* before the child is sent off to the 'shop', so that he has to retain the three (or four) items in his memory. Encourage the children to repeat their instructions on the way to the shop. Also encourage them to pick up the items in the order they were given. Children who can cope with three items can have their shopping lists increased to four.

Level I

Level II

Level III

Level IV

Circle Time

Hall/PE

Literacy

Topic Work

Drama

Small Group

© C Delamain &
J Spring 2000.
Photocopiable

LANGUAGE AND LITERACY

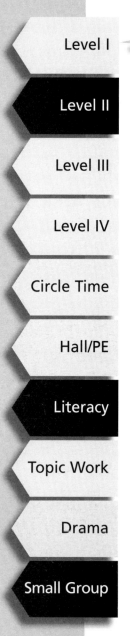

Matching Pictures

Aim
To be able to remember a sequence of three named pictures and reproduce the sequence accurately.

 **Equipment**
A set of paired pictures.
A screen or barrier to be placed between you and the child. (This can be a box, a box lid propped up, or anything that screens what you are doing from the child's sight, and can be easily removed and replaced.)

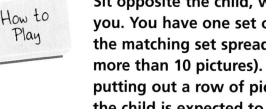 **How to Play**
Sit opposite the child, with the screen between you. You have one set of pictures, the child has the matching set spread out in front of him, (not more than 10 pictures). Explain that you will be putting out a row of pictures in front of you, and the child is expected to produce an exact copy. Say, 'I'm putting down a chair, a dog, and a house' (suiting the action to the words). When the child has made his selection, the screen is lifted and the child can see if he has matched your sequence. Continue round the group until everyone has had enough turns.

 **Tip**
You must remember to put your pictures down from *right* to *left*, to match the sequence of the child, who is always shown to put his pictures down from *left* to *right*.

 **Extension**
Increase the number of pictures to each child's limit.

© C Delamain & J Spring 2000. Photocopiable

LANGUAGE AND LITERACY

Chinese Whispers (i)

Aim
To be able to repeat a simple sentence.

None.

The children stand in a line, one behind the other. Explain that you are going to pass a message along the line. Each child must whisper the message in turn to the child behind until it reaches the end of the line. The last child must carry out the action contained in the message.

Come and tell me your name
Come and show me your shoes
Go and touch the door

Level I

Level II

Level III

Level IV

Circle Time

Hall/PE

Literacy

Topic Work

Drama

Small Group

© C Delamain &
J Spring 2000.
Photocopiable

LANGUAGE AND LITERACY

Auditory Memory

Colour Jumping

Aim
To be able to remember a sequence of two or three colours.

Hoops, as many different colours as possible.

Lay the hoops out on the floor touching each other and forming a large circle. Choose the first child to have a turn. Tell the child to jump into a yellow and then a blue hoop. The next child might be told to jump into a green and then a blue hoop. The easiest instruction to follow simply names the colours ('red, blue'). It is harder if you use more language, for example 'Jump into a yellow hoop, then a red one'. The complexity of the instruction can be matched to each child's ability. Children who can remember two colours can be promoted to remembering three.

If enough hoops are available, you can make this game more fun by having a large lay-out of hoops, and starting two or more children off at the same time.

© C Delamain &
J Spring 2000.
Photocopiable

PAGE 192   DEVELOPING BASELINE COMMUNICATION SKILLS

LANGUAGE AND LITERACY

Magic Carpets

Aim

To be able to remember when you hear your name as one of a group.

Skipping ropes or large hoops.

Mark out several circles on the floor using skipping ropes, hoops or chalk. The circles should be large enough for up to four children to stand inside. Explain that these are magic carpets, and each one will take you to a different place. The children stand in a group in the middle of the room. Stand by the first circle, and call out three names ('Christopher, Mary and Abigail, come to this magic carpet'). When the children arrive, use your imagination to tell them where the carpet is going, and tell them something interesting about it, inviting a little discussion. Then move to the next circle and repeat the process with the next three children, but of course naming a different destination for their magic carpet. Continue until all the children are on a carpet.

Increase to calling out four names at a time.

© C Delamain &
J Spring 2000.
Photocopiable

LANGUAGE AND LITERACY

- Level I
- Level II
- **Level III**
- Level IV
- Circle Time
- **Hall/PE**
- Literacy
- Topic Work
- Drama
- Small Group

Pets' Corner

Aim
To be able to remember when an item is eliminated from a list.

 None (unless you would like to have a list of animals with their preferred foods, as a reminder).

 **Divide the children into two teams standing facing each other. The members of one team are the animals in Pets' Corner at the zoo, the members of the other team are the keepers. Each keeper is responsible for the animal standing opposite him. Say to the first keeper, 'Your animal (say you choose a horse) likes to eat grass, apples and sugar lumps. The zoo has run out of apples. What can you give him to eat?' The keeper names one of the remaining items, and takes his 'animal' off to the side of the room. Continue down the line.**

Donkey – carrots, bread, hay
Monkey – bananas, nuts, grubs
Rabbit – lettuce, carrots, cabbage
Dog – meat, bones, biscuits
Cat – fish, milk, cat food

© C Delamain &
J Spring 2000.
Photocopiable

LANGUAGE AND LITERACY

Whose News?

Aim
To be able to remember someone else's news for a few minutes, and re-tell it.

Level I

Level II

Level III

Level IV

Bowl or box containing counters, two of each colour, enough altogether for each member of the group to have one.

Circle Time

The children sit in a semi-circle in front of you. The bowl of counters is passed round, each child taking one, so that by the end the children are paired off by colour. The children identify their colour partners, and move to sit next to each other. The first child in the group (child A) tells his news. His colour partner (child B) is asked to repeat child A's news in his own words. If he cannot remember, encourage child A to give him a clue. If he gets it wrong child A should be given the opportunity to correct him. Then the turn moves on to the second pair of colour partners, and so on round the group.

Hall/PE

Literacy

Topic Work

Drama

Small Group

This game is best played with small enough groups to permit a re-shuffle of partners half-way through, allowing every child to tell his own news as well as relaying his partner's news.

© C Delamain &
J Spring 2000.
Photocopiable

DEVELOPING BASELINE COMMUNICATION SKILLS

LANGUAGE AND LITERACY

Chinese Whispers (ii)

Aim
To be able to repeat a more difficult sentence.

None.

This is a more difficult version of the Chinese Whispers **played at** *Level II*.

The sentences to be passed on can contain three different ideas. They do not have to refer to the 'here and now', and can be about any subject. Children stand in a line, one behind the other. Explain that you are going to pass a message along the line. Each child must turn and whisper the message to the child behind him until it reaches the end of the line. The last child must come and relay the message to you.

It was *cold* and *windy* on *Thursday*
The *man* on *skates* fell over
Rabbits in *cages* need *water*
The *snowman melted* in the *sun*
Does the *bus stop* at *your house*?
The *plane* made *patterns* in the *sky*

© C Delamain &
J Spring 2000.
Photocopiable

PAGE 196   DEVELOPING BASELINE COMMUNICATION SKILLS

LANGUAGE AND LITERACY

Ring me Up!

Aim
To be able to remember and repeat a string of three or four digits.

Two toy telephones.

The children stand in two lines facing each other. The first child in line 1 (child A) is given one telephone, the first child in line 2 (child B) is given the other. Explain that child A is going to telephone child B and invite him to a party. Whisper a string of three numbers into child A's ear. He must call them out loudly, and child B must acknowledge by picking his receiver up and repeating 'his' number. He then joins child A in the 'party' group, and the telephones are passed on to the next pair of children.

Start with three numbers at a time only until confidence is built. The memory load is made greater if the numbers are said slowly. At the four-digit level, children who are struggling can be helped if you group the numbers as you say them at first, thus '2 ,3 . . . 7, 5'.

- Level I
- Level II
- **Level III**
- Level IV
- Circle Time
- **Hall/PE**
- Literacy
- Topic Work
- Drama
- **Small Group**

© C Delamain & J Spring 2000. Photocopiable

DEVELOPING BASELINE COMMUNICATION SKILLS

LANGUAGE AND LITERACY

Postbag (ii)

Aim
To be able to remember a series of items and add to it without visual support.

None.

The children sit in a circle. Explain that you are going to help the postman remember what he has got in his bag. Start the game off, saying 'In my bag I've got a . . .' and name an item. The first child in the group must repeat 'In my bag I've got a . . .' (repeating the item chosen by you) 'and a . . .', adding an item of his own. The next child must remember the first two items and add one of his own, and so on round the group. The game stops when the list becomes too long for anyone to remember.

© C Delamain &
J Spring 2000.
Photocopiable

LANGUAGE AND LITERACY

Parrots

Aim
To be able to repeat a 'nonsense' utterance.

List of 'nonsense' utterances.

The children sit in a circle around you. Explain that you want to buy a parrot, and you are looking for one that talks really well and can copy what you say. (It may be necessary to explain to some of the children that some parrots can talk, and that they imitate what they hear.) The children are all parrots, and you are going to test them out to see which one to buy. If this game is felt to be too discriminatory, divide the children into very small groups, and arrange to buy the three or four parrots in the best performing group! You then choose 'parrots' in turn and asks them to repeat the 'nonsense' words.

Bangerwallop, Widdershins, Spanderwitch, Pifflepuff, Abracadabra, Splatterposh, Bumbletump, Tragglepoop

You can invent nonsense words by taking a real word and changing the vowel sounds (eg, crocodile becomes cricodole, elephant becomes olophint).

Invent little nonsense phrases, such as 'Piffle my wudgeon'.

- Level I
- Level II
- Level III
- **Level IV**
- **Circle Time**
- Hall/PE
- Literacy
- Topic Work
- Drama
- **Small Group**

© C Delamain & J Spring 2000.
Photocopiable

DEVELOPING BASELINE COMMUNICATION SKILLS

LANGUAGE AND LITERACY

AUDITORY MEMORY

Pantomimes

Aim
To be able to repeat sentences heard in a story.

Equipment
Short stories (photocopiable stories available in Activity Resources, pp268–270).

How to Play
The children sit in front of you separated into two groups. Explain that you are going to read them a story, but that you will need some help, as you are feeling rather forgetful. When you 'forget' what somebody has said, you will point to one or other of the groups and ask them to remind you 'What did he say?'

Tip
This can get quite noisy!

Level I
Level II
Level III
Level IV
Circle Time
Hall/PE
Literacy
Topic Work
Drama
Small Group

© C Delamain &
J Spring 2000.
Photocopiable

LANGUAGE AND LITERACY

Fact Finder

Aim
To be able to remember specific facts heard in the course of a short story.

Very short stories (photocopiable stories available in Activity Resources, pp271–273).

The children sit in a semi-circle in front of you, in pairs or threes. Explain that you are going to read a little story to them. Brief them as to what the story is broadly about. Then allocate to each pair or group specific facts to remember. When the story ends, ask the pairs or groups in turn to remind you of the facts that they were asked to focus on. At the end you might like to re-tell the story so the children can see if they were right.

Group A – remember the name of the character
Group B – remember how old he was
Group C – remember where he was going

Level I

Level II

Level III

Level IV

Circle Time

Hall/PE

Literacy

Topic Work

Drama

Small Group

© C Delamain &
J Spring 2000.
Photocopiable

LANGUAGE AND LITERACY

Matching Pictures (ii)

Aim
To be able to remember a sequence of four named pictures and reproduce the sequence accurately.

A set of paired matching pictures.
A screen or barrier to be placed between you and the child. (This can be a box, a box lid propped up, or anything that screens what you are doing from the child's sight, and can be easily removed and replaced.)

Sit opposite the child, with the screen between you. You have one set of pictures, the child has the matching set spread out in front of him (not more than 10 pictures). Explain that you will be putting out a row of pictures in front of you, and the child is expected to produce an exact copy.

Say 'I'm putting down a chair, a dog, a house and a tree (suiting the action to the words). When the child has made his selection, the screen is lifted and the child can see if he has matched your sequence. Continue round the group until everyone has had enough turns.

You must remember to put your pictures down from *right* to *left*, to match the sequence of the child, who is always shown to put his pictures down from *left* to *right*.

© C Delamain &
J Spring 2000.
Photocopiable

LANGUAGE AND LITERACY

Jack and Jill

Aim
To be able to carry out a simple instruction involving four ideas.

A worksheet for each child and one for you, depicting Jack and Jill (photocopiable pictures available in Activity Resources p274).
List of instructions for your use (available in Activity Resources p275).

The children and you all have a worksheet, and a plentiful supply of crayons. The children sit round the table, you sit a little way away so that the children cannot see your worksheet. You have a list of instructions. Explain to the children that you are going to tell them different things to add to their pictures, and that you are going to add to yours at the same time. Then begin to tell them what to do. When you reach the end, compare your picture with those of the children. Whose is the nearest to yours?

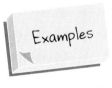

Put a red dot on Jack's hat.
Put a blue cross on Jill's nose.

Because copying is possible when the game is played in this way, it is worth trying it out individually with any children who seem unsure of what they are doing.

Level I

Level II

Level III

Level IV

Circle Time

Hall/PE

Literacy

Topic Work

Drama

Small Group

© C Delamain &
J Spring 2000.
Photocopiable

Developing Baseline Communication Skills
LANGUAGE AND LITERACY

PHONOLOGICAL AWARENESS

Level I

206 / Clap Your Name
207 / Switch Me Off
208 / Pass on the Code
209 / Poems Please!
210 / Hurry them Up!

Level II

211 / Come When I Clap
212 / My Mistake
213 / Beanbags
214 / Incy-Wincy-Mincy-Pincy
215 / Pitter-Patter

Level III

216 / On the Beat
217 / Cats and Rats
218 / Sound Ladders
219 / Rhyming Families
220 / Stepping Stones

Level IV

221 / Prove It
222 / Rhyme the Number
223 / TV Tongue Twisters
224 / Odd One Out
225 / Bob's Bunkbed

LANGUAGE AND LITERACY

PHONOLOGICAL AWARENESS

Clap Your Name

Aim
To understand that there are differing numbers of beats (syllables) in names.

None.

The children sit in a semi-circle in front of you. Explain that you are going to clap everybody's name. Ask the first child to stand up. You say his name, synchronising a clap or claps with your speech, ('John' – one clap as you say the name, 'Be-cky' – two claps as you say her name). The clap(s) must be exactly simultaneous with the name, not before or after. Then ask the whole group to say and clap the name. Proceed round the group until everyone's name has been said and clapped both by you and by the group.

This is sometimes a very difficult task for children to master, and may need repeating several times before the children are ready to move on to harder syllabification tasks.

Vary it by stamping the names instead of clapping, or by passing a drum round the group so that each child can beat out the syllables in his own name as he says it.

Use the form of name that the children are most comfortable with (eg, Chris rather than Christopher, if that is what he is usually called, even though it is tempting to keep Christopher for a three-syllable name!).

© C Delamain & J Spring 2000. Photocopiable

LANGUAGE AND LITERACY

Switch Me Off

Aim
To be able to supply the missing rhyming words in familiar nursery rhymes and songs.

None.

The children sit in a semi-circle in front of you. Explain to the children that you are going to say or sing some nursery rhymes and songs with them, but that your loudspeaker has something wrong with it, and sometimes it will switch your voice off. The children will have to supply the missing word for you until your speaker comes on again. Choose a nursery rhyme or song with which the children are familiar. Say or sing it once. Then say or sing it again, but this time 'switch your voice off' on the rhyming words, (Humpty Dumpty sat on the wall, Humpty Dumpty had a great . . ., All the King's horses and all the King's men, Couldn't put Humpty together . . .).

Good songs and rhymes to choose will have clear and unambiguous rhyming words, rather than doubtful rhymes such as 'water' and 'after' in Jack and Jill.

'Humpty Dumpty', 'Little Jack Horner', 'Little Boy Blue', 'Mary Mary Quite Contrary', 'Incy Wincy Spider', 'Ring-a-ring o' Roses', 'Row, Row your Boat', 'Sing a Song of Sixpence'. (NB The final rhyme in the last one is a bit dubious – 'nose' and 'clothes'! It is important to explain such ambiguities to children so that they learn what is really a rhyme and what is only approximately a rhyme.)

Level I

Level II

Level III

Level IV

Circle Time

Hall/PE

Literacy

Topic Work

Drama

Small Group

© C Delamain &
J Spring 2000.
Photocopiable

LANGUAGE AND LITERACY

PHONOLOGICAL AWARENESS

Pass on the Code

Aim
To be able to imitate one, two or three claps.

None, but this game is better if there are two or three adults available.

The children and adults sit in a circle, with the adults seated well apart from each other. Explain that you are going to give a coded message, by clapping, to your next door neighbour, who must pass it on to his next door neighbour, and so on round the circle. When the message reaches another adult, either change the number of claps and send a new message on its way, or turn round and send the message back in the opposite direction.

Clap rhythms (*clap* – clap clap)
Clap slowly (clap ...... clap)
Clap fast (clapclapclap)
Clap a mixture of the above

© C Delamain &
J Spring 2000.
Photocopiable

LANGUAGE AND LITERACY

Poems Please!

Aim
To be able to remember a pair of rhyming words.

An assortment of pictures with easily-rhymed names, one for each child.
A bag.

This game requires you to become a minor poet! The children sit in a circle round you. Explain that the children are all going to be given their very own poem, which they are to try and remember. Place the pictures in the bag. The first child comes up and draws a picture out of the bag. Improvise a rhyme using the name of the picture. The child returns to his place taking his picture with him. Continue round the circle. At the end, ask the children to repeat their 'poems'. How many can remember them?

A *cat* in a hat
A *fish* on a dish
A *frog* on a log
A *rat* on a mat
A *bee* in a tree
A *car* stuck in tar
A *man* in a van
A *dolly* with a lolly

Level I

Level II

Level III

Level IV

Circle Time

Hall/PE

Literacy

Topic Work

Drama

Small Group

© C Delamain &
J Spring 2000.
Photocopiable

DEVELOPING BASELINE COMMUNICATION SKILLS

LANGUAGE AND LITERACY

PHONOLOGICAL AWARENESS

Hurry them Up!

Aim
To be able to repeat words increasingly fast.

None.

Divide the class into two groups. Explain that one group are animals, and the other group are the farmers driving the animals along. The farmers are in a hurry, and the animals are being dreadfully slow. Give the animal group an animal sound to make (choose from moo, quack, honk, cluck, baa, neigh). Tell them that when you say 'Go!' they are all to start making their sound and keep saying it faster and faster until you say 'Stop!'. At the same time the farmers are to urge them on by saying 'Faster-faster-faster' as quickly as they can until you say 'Stop!'. Change over groups. Give the new animal group a different sound to make, and give the new farmer group a different word ('Quicker! Quicker!' or 'Hurry! Hurry!'). Set them off with 'Go!' again, and let them keep it up until you say 'Stop!'.

This is a quick two- or three-minute warm-up activity or one to use to release pent-up energy. It can get quite noisy, so is a good game to play outside.

© C Delamain &
J Spring 2000.
Photocopiable

DEVELOPING BASELINE COMMUNICATION SKILLS

LANGUAGE AND LITERACY

Come When I Clap

Aim
To be able to recognise the number of syllables in your own name.

Ten large coloured paper circles or balloons.

The children are going to be asked to form groups according to the number of syllables in their names. Use the coloured circles or balloons to indicate where the various groups are to collect; one circle = single syllable, two circles = 2 syllables, three = 3 syllables, four = 4 syllables. Explain that you are going to 'call' the children by clapping their names. Go to the one-syllable area, give a single clap, and await the arrival of the children with single-syllable names. Then move on to the two-syllables areas, and repeat the process, until all the children are collected in their 'syllable' families.

This is an area where errorless learning is important, so help and prompting should be given to children who make a mistake or are unsure.

- Level I
- **Level II**
- Level III
- Level IV
- Circle Time
- **Hall/PE**
- **Literacy**
- Topic Work
- Drama
- Small Group

© C Delamain &
J Spring 2000.
Photocopiable

LANGUAGE AND LITERACY

PHONOLOGICAL AWARENESS

My Mistake

Aim
To be able to detect when rhyming words are replaced by non-rhymes in familiar nursery rhymes and songs.

 None.

 **The children sit in a semi-circle in front of you. Explain that today you are feeling rather forgetful, and may well make some mistakes as you say or sing the rhymes. Ask the children to call 'stop' when they hear a mistake. Can they put you right? Say or sing the rhymes, changing the rhyming words for words of similar meaning but that do not rhyme.**

 Humpty Dumpty sat on the wall, Humpty Dumpty had a great *tumble*.
All the king's horses and all the king's men,
Couldn't put Humpty together *any more*.

Little Miss Muffet
Sat on a *cushion*
Eating her curds and whey
There came a big spider
And sat down *next to her*
And frightened Miss Muffet *off*.

© C Delamain &
J Spring 2000.
Photocopiable

DEVELOPING BASELINE COMMUNICATION SKILLS

LANGUAGE AND LITERACY

PHONOLOGICAL AWARENESS

Beanbags

Aim
To be able to match actions to the number of syllables in words.

A beanbag.

The children stand in a circle, far enough apart to be able to throw the beanbag from one to another. You stand in the middle, and give the beanbag to a child (child A). Explain that you will be saying a word, if the word has one syllable, the beanbag is to be thrown on to child A's next-door neighbour (child B) only. If the word has two syllables, child A throws the bag to child B, and child B throws it on to the next child (child C). Three-syllable words mean that the beanbag must be thrown on three times.

If difficulties with catching and dropping the beanbag interfere with the aim of the game, stand the children a little closer together, and let them pass the beanbag rather than throwing it.

Level I

Level II

Level III

Level IV

Circle Time

Hall/PE

Literacy

Topic Work

Drama

Small Group

© C Delamain &
J Spring 2000.
Photocopiable

DEVELOPING BASELINE COMMUNICATION SKILLS

PAGE 213

LANGUAGE AND LITERACY

PHONOLOGICAL AWARENESS

Incy-Wincy-Mincy-Pincy

Aim
To be able to generate a string of rhyming words and non-words by changing the initial sound.

A watch with a second hand.

The children sit in a circle round you. Explain that you are going to make up a string of real and nonsense words. Give a demonstration, 'Incy-Wincy-Mincy-Pincy-Lincy-Bincy', etc. You will start off with a word or nonsense word, and each child will add a word or non-word to the list. You will time how long each 'string' can be kept going. The object of the game is to keep a string going for the longest possible time. At the end of a game, make a note of the longest time, and see if the children can beat it when you play again.

It is really important to emphasise to the children that nonsense words are absolutely acceptable.

Once the children are finding this game quite easy, introduce an element of competition by having two teams. Give the teams a word alternately, and see which team can keep going the longest.

© C Delamain & J Spring 2000. Photocopiable

LANGUAGE AND LITERACY

PHONOLOGICAL AWARENESS

Pitter-Patter

Aim
To be able to say words which need rapid shifts from one speech sound to another.

Word lists available in Activity Resources, p276.

Divide the children into three or four groups, spread out into different parts of the room. Tell each group what they are to be (eg, group 1 raindrops, group 2 rabbits, group 3 hares, group 4 kangaroos). Each group has its own sound. Raindrops say 'pitter-patter', rabbits say 'hippety-hoppety', hares say 'lippety-loppety', kangaroos say 'humpety-jumpety'. Get each group to practise its sound. On the command 'Go!', the groups start moving quietly around the floor, saying their sounds softly as they go. The groups will mingle, but the children should continue saying their group sound softly as they move about. After a few minutes, tell the children to re-form into their groups. How many of them can still remember and say their sound? See if the groups can chant their sound together faster and faster.

- Level I
- **Level II**
- Level III
- Level IV
- Circle Time
- **Hall/PE**
- **Literacy**
- Topic Work
- Drama
- Small Group

© C Delamain &
J Spring 2000.
Photocopiable

DEVELOPING BASELINE COMMUNICATION SKILLS

LANGUAGE AND LITERACY

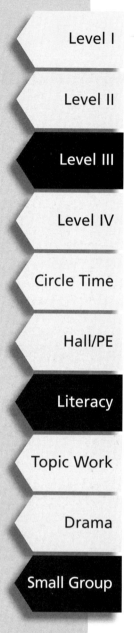

Phonological Awareness

On the Beat

Aim
To be able to differentiate words by the number of syllables they contain.

Three boxes (shoe box size or larger) numbered 1, 2 and 3, and also marked with dots to represent numbers. A collection of small one-, two- or three-syllable objects or pictures (photocopiable list available in Activity Resources, pp277–278).

The children sit in a circle. The boxes are placed in the middle. Explain that today you cannot always talk, and you are going to ask for what you want by sending clapped messages. Put two objects or pictures containing different numbers of syllables in front of the first child. Name both in turn, clapping the syllables at the same time. Then clap the one you want *without speaking*. Can the child select the right one? The object or picture is placed in the appropriate box. Move on to the next child, and present another choice, and so on round the circle.

To ensure confidence, begin by presenting a one-syllable and three-syllables choice for maximum contrast. As the children's skill increases, make the contrast harder by including one-syllable and two-syllables, or two-syllables and three-syllables choices.

When children make mistakes, correct and help them choose the right one, reinforcing the choice by clapping the word again. Errorless learning is important at this stage.

© C Delamain & J Spring 2000. Photocopiable

LANGUAGE AND LITERACY

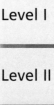

PHONOLOGICAL AWARENESS

Cats and Rats

Aim
To be able to select two rhyming words out of a choice of three.

Equipment

Sets of three pictures. In each set two words rhyme, the third is the odd one out.
A container.

How to Play

Children sit in a circle round you. You have the pictures and the container. Explain that the children are going to find their rhyming partners. Put the first set of three pictures in the container, and call out the first three children from the circle. Each child takes a picture from the container, and they must decide between them which two are the rhyming partners. These two then return to sit next to each other, and the first set of three pictures is removed from play. The child who was the 'odd one out' waits by you. Put the next set of three pictures into the container, and call out two more children. These three pick a card each, and identify the rhyming partners as before. Proceed round the circle until everyone is paired off. One child will be left over at the end and ingenuity may be needed to ensure that he does not feel he has failed.

Tip

Give as much help as necessary to ensure success. Some children may have internalised the concept of rhyme sufficiently to be able to identify the rhyming pairs simply by seeing the pictures. Others will still need you to name the pictures. Avoid choosing words which start with the same sound, as children can easily become confused between 'starts with the same sound' (alliteration) and rhyme.

- Level I
- Level II
- **Level III**
- Level IV
- **Circle Time**
- Hall/PE
- **Literacy**
- Topic Work
- Drama
- **Small Group**

© C Delamain &
J Spring 2000.
Photocopiable

LANGUAGE AND LITERACY

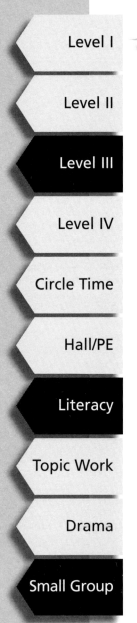

Sound Ladders

Aim
To be able to imitate letter sounds, relying on listening only.

A picture of a ladder, for the group to share. The rungs should be wide enough apart for counters to be put on them.
Counters.

This game can be played as soon as the children have been introduced to the idea that words are made up of sounds, and have practised making the sounds themselves. Seat the children round the table, with the ladder and counters in the middle. Explain that you are going to turn your back and make the sound of a letter. This is to ensure that the children do not lip read the sound, but rely on their listening skills. The first child in the group must copy your sound. If he is correct, he puts a counter on the first rung of the ladder. Proceed to the next child and repeat the process. If a child makes a mistake, a counter is taken off the ladder. The target is to get a counter on every rung of the ladder.

If the children are poor listeners, start by making sounds that are quite distinct from one another, for example 'b' and 's'. As the children's skill increases, you can move on to sounds that are more similar, and ultimately to sounds that can be hard to distinguish, for example 's' and 'f'.

(IMPORTANT: Please see notes in 'How to Use This Book' on the production of letter sounds.)

© C Delamain &
J Spring 2000.
Photocopiable

DEVELOPING BASELINE COMMUNICATION SKILLS

LANGUAGE AND LITERACY

Rhyming Families

Aim
To be able to sort words by their sound into rhyming groups.

As many sets of rhyming pictures as possible, at least three in each set (cat/hat/mat/ – dog/frog/log/ – pig/wig/fig – man/van/pan – case/face/race).
Two boxes.

The children sit in a circle round you. You have the boxes nearby, and the picture sets at hand. Explain that you are going to sort out families of rhyming words. One family lives in each box. Call out two children, who each stand beside a box. Place a key picture beside each box – for example 'cat' by one box, and 'pig' by the other. Make sure each child knows the word he is trying to rhyme. Show and name the remaining pictures of the rhyming groups one by one in random order. Encourage the children to identify the ones that belong to 'their' family, and put them into the family box. Repeat with further pairs of children and more sets of rhyming words.

When helping or explaining, refer to the 'at' box and the 'ig' box, as well as the 'cat' box and the 'pig' box, to encourage the children to attend to the stem of the word.

When selecting sets of rhyming words, try to avoid using pictures which the children would normally refer to by a different name. (For example, do not be tempted to include 'hog' in the 'og' group, as this would normally be called 'pig'.) However it is fine to include some words which are new to them.

Level I

Level II

Level III

Level IV

Circle Time

Hall/PE

Literacy

Topic Work

Drama

Small Group

© C Delamain &
J Spring 2000.
Photocopiable

LANGUAGE AND LITERACY

Phonological Awareness

Stepping Stones

Aim
To be able to make the sounds of the consonants (phonemes) while stepping on the written letters (graphemes).

Equipment

Large pieces of paper or card, with the lower case consonants written on them in clear bright colours, (several of each consonant).
Blu-Tack®.

How to Play

Select the sound you choose to work on (eg, 'p'). Place five or six 'p' cards on the floor, close enough to each other for the children to step or jump from one to the other. Arrange them so that they form a line of stepping stones between two chairs or two benches. Line the children up at the beginning of the stepping stones. Explain that they have to cross the river by jumping or stepping on each stone, without getting their feet wet. At the same time they must say the sounds written on the stones. Anyone falling off a stone or missing out a sound, will have to be rescued, taken back to the bank, and sent off again. You can embellish this game if you wish by having the river infested by sharks, or the children pursued by crocodiles.

Tip

Fix the stepping stones down with Blu-Tack® so that they do not move about when the children step on them. This is particularly important in the case of letters such as 'p' and 'b' which could become reversed.

Extension

Include two or more different sounds on the stepping stones. The children must get the sounds right or go back to the bank.

© C Delamain &
J Spring 2000.
Photocopiable

LANGUAGE AND LITERACY

Prove It

Aim
To be able to group words according to the number of syllables they contain.

Three boxes (shoe-box size or larger) numbered 1, 2 and 3. Also mark them with one, two and three dots. A collection of one-, two- and three-syllables objects or pictures, equal numbers of each (list and pictures available in Activity Resources pp277–278).

The children sit in three groups round you. Each group has a box, you have objects or pictures ready to hand. Explain that the groups are going to 'claim' the objects or pictures that belong in their box, and prove their claim by clapping the syllables. Go round each group first, saying and clapping a word which would belong in their box. You then select an object or picture, say its name, (*no clapping*) and ask the groups in turn whether they think it belongs to them. Allow for some debate within the group until they arrive at a consensus. If they decide to lay claim to the object, they must prove it by clapping the syllables. Correctly claimed items go into the group's box; incorrectly claimed items go back into the central pile. The winning group is the one who has accumulated the most items at the end.

Level I

Level II

Level III

Level IV

Circle Time

Hall/PE

Literacy

Topic Work

Drama

Small Group

© C Delamain &
J Spring 2000.
Photocopiable

LANGUAGE AND LITERACY

Rhyme the Number

Aim
To be able to think of a word that rhymes with the number thrown on the die.

Giant die.

The children sit in a circle round you. The die is given to the first child, who rolls it. You name the number thrown, and the child must think of a word which rhymes with that number ('four – door', 'three – tree', 'two – shoe'). The die then passes to the next child in the circle.

If a child is stuck for a word, give a clue ('Two – think of something you wear on your feet'). If the child still produces a non-rhyme, give a further clue by offering a choice ('Could that be sock or shoe?'). Children who are persistently unable to generate a rhyme should go back a stage to practising rhyme recognition. Five is a hard one to rhyme.

If a child is managing this game easily, you should stop saying the number for him. He may either say it for himself, or may be able to produce a rhyme without having had to hear the number or say it out loud. This is the internalisation of the rhyming process.

© C Delamain &
J Spring 2000.
Photocopiable

LANGUAGE AND LITERACY

TV Tongue Twisters

Aim
To be able to use rapid and precise speech in some sentences and rhymes.

Level I

Level II

Level III

Level IV

Circle Time

Hall/PE

Literacy

Topic Work

Drama

Small Group

Equipment

List of tongue twisters (photocopiable list available in Activity Resources, p279).
Guitar or piano.

How to Play

This game can be played as part of a music lesson. Divide the children into two groups. Explain that they are going to take turns to practise saying some tricky sentences. Actors and announcers on television have to learn to do this too. Group A will practise one first, and when group B think they are good enough to say it on television they must give them a round of applause. Then it will be group B's turn to rehearse a sentence.

Use the piano or guitar to accompany the children, give them a slow steady pace, and mark the syllables. Gradually speed up a little, and then stop the musical accompaniment.

Examples

Sa- mmy- snake- slipped- off - to- sleep
Pe - ter Pi - per - picked - some - pe - pper

Tip

You can use clapping instead of music to help the children.

Extension

Go on to practising action songs and rhymes at increasing speed (for example, 'Head, Shoulders, Knees and Toes').

© C Delamain & J Spring 2000.
Photocopiable

DEVELOPING BASELINE COMMUNICATION SKILLS

LANGUAGE AND LITERACY

PHONOLOGICAL AWARENESS

Odd One Out

Aim
To be able to detect the odd one out in a collection of words which all start with the same sound.

 Collections of small items beginning with the same sound and some that have different beginnings. A bag.

 **Put the objects into the bag. The children sit in a semi-circle in front of you.**

Explain that there are some things in your bag which all begin with the same sound, but one thing got into the bag by mistake. Can the children spot the odd one out? If they can, they must call 'Odd one out'. The children come up in turn and take an object out of the bag, saying its name as they do so. When they have identified the odd one out, repeat with another sound collection.

 A 'b' collection might include: book, beanbag, box, bead, bed, (from doll's house), biscuit (from toy food pack), bowl, with the odd one out being a sausage.

 To help ensure success in the early stages, choose words in which the initial consonant is followed by a vowel. As the children gain in skill and confidence, introduce words which begin with more than one consonant (block, bracelet).

© C Delamain &
J Spring 2000.
Photocopiable

DEVELOPING BASELINE COMMUNICATION SKILLS

LANGUAGE AND LITERACY

PHONOLOGICAL AWARENESS

Bob's Bunkbed

Aim
To be able to think of words that begin with the same sound.

Equipment

None.

How to Play

The children sit in a circle around you. Explain that you are going to plan what to put in the bedroom for a boy called Bob. Its quite a big bedroom, so you can put in lots of things, but they must all begin with 'b' as that is the sound that begins Bob's name. Use the letter name as well as the sound. Start the game off by suggesting a bunkbed. Then go round the circle asking for contributions from the children. Repeat with other's children's names and different rooms. Allow silly suggestions so long as they begin with the right sound!

Level I

Level II

Level III

Level IV

Circle Time

Hall/PE

Literacy

Topic Work

Drama

Small Group

© C Delamain &
J Spring 2000.
Photocopiable

DEVELOPING BASELINE COMMUNICATION SKILLS

Developing Baseline Communication Skills
ACTIVITY RESOURCES

Personal and Social Development

228 / The Farmer Wants a Horse
229–230 / Build It
231–232 / Mime Story
233 / Who is Asleep?
234–235 / Lions and Tigers
236 / Scavenge Hunt
237–240 / Happy–Sad
241–244 / Angry–Scared
245 / All Change
246–251 / Spin-a-Word

Language and Literacy

252 / Where's Granny Going?
253–255 / It's a Funny World
256–257 / Work it Out!
258 / Big Green Apples *and* Only One Left
259 / Colour Families
260–262 / George the Giant
263–264 / Spot the Difference
265–267 / Zig at the Zoo
268–270 / Pantomimes
271–273 / Fact Finder
274–275 / Jack and Jill
276 / Pitter Patter
277–278 / On the Beat *and* Prove It
279 / TV Tongue Twisters

PERSONAL AND SOCIAL DEVELOPMENT

Level I

Turn Taking

The Farmer Wants a Horse

The farmer wants a horse,
The farmer wants a horse,
Ee-aye ee-aye,
The farmer wants a horse.

The horse wants a goat,
The horse wants a goat,
Ee-aye ee-aye,
The horse wants a goat.

The goat wants a pig,
The goat wants a pig,
Ee-aye ee-aye,
The goat wants a pig.

The pig wants a cat,
The pig wants a cat,
Ee-aye ee-aye,
The pig wants a cat.

The cat wants a mouse,
The cat wants a mouse,
Ee-aye ee-aye,
The cat wants a mouse.

The mouse wants some cheese,
The mouse wants some cheese,
Ee-aye ee-aye,
The mouse wants some cheese.

Build It

Level III

Turn Taking

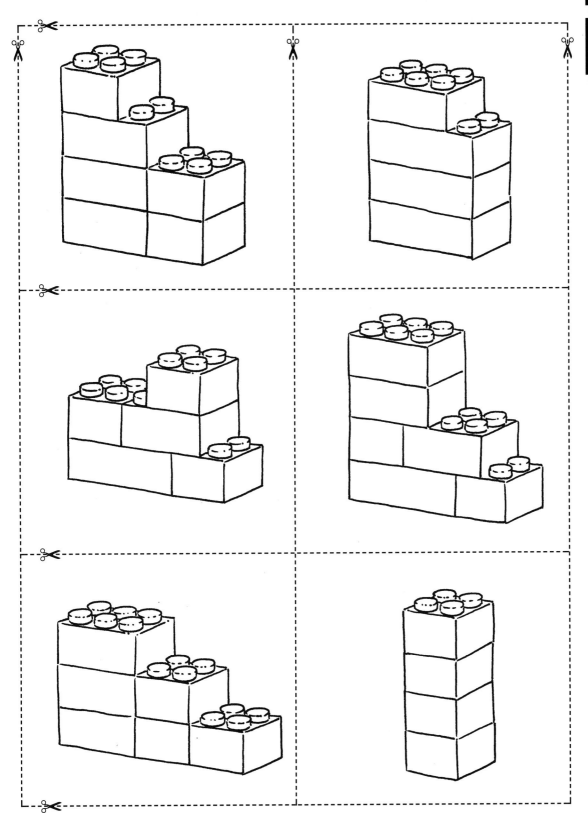

These pictures have to be coloured before use.

Build It *(continued)*

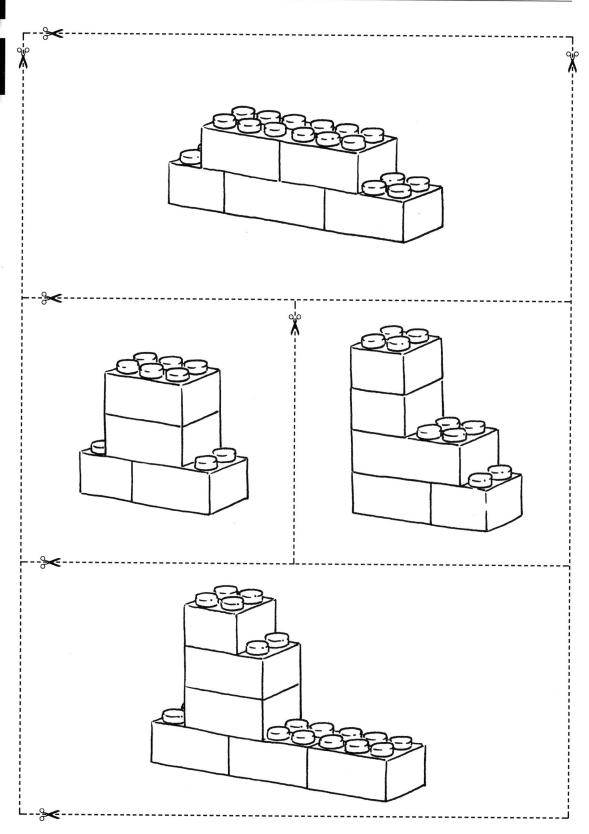

These pictures have to be coloured before use.

Mime Story

Level IV

Body Language

1 Ted's Bad Mood

Parentheses indicate where the reader should pause for a mime.

Ted was in a very bad mood. He had to go shopping with his mum, which meant lots of waiting about while she chose boring things like potatoes and soap. Then it meant helping to carry it all, and packing it away when they got home. Ted made a horrible face at himself in the mirror. ( ) It made him feel a bit better, but not much. He pulled off his pyjamas and threw them into the corner on the floor. ( ) When he was dressed, he brushed his teeth ( ) and washed his face. ( ) Then he put his comb under the tap ( ) so he could wet his hair when he combed it. ( ) He pulled a bit of his hair so it stuck straight up ( ) because his mum said it made him look like a bottle brush, and she didn't like it. Ted was in *that* sort of a bad mood.

Things got a bit better on the way to the shops. Ted saw a friend of his, and gave him a wave ( ) and then stuck his tongue out at him. ( ) It began to rain a bit, so he could put up his brand new red umbrella. ( ) When they got into the shop, things *really* began to look up. There was a colouring competition on, and if you won it, you got a brilliant computer game. A man was handing out the pictures. Ted reached out for one, ( ) and gave the man his best smile. ( ) The man shook Ted's hand ( ) and wished him luck. Ted couldn't wait to get home and start colouring. There was a big stand of the computer games near the check-out and Ted pointed them out to his mum. ( )Then he helped her load heaps of things into the trolley ( ), pushed it to the check-out for her ( ) handed the money to the check-out lady ( ) and hurried his mum along the road so fast her feet hardly touched the ground. When they reached their front door, Ted put the key in the lock, ( ) turned it, ( ) and pushed open the door. ( ) Then he quickly got out his colours, and began to colour his picture. ( )

You will have to wait till another day to find out if Ted won a computer game.

Mime Story *(continued)*

2 Ted's Surprise

Do you remember the story about Ted and the colouring competition? He finished colouring his picture, put it in an envelope, ( ) licked a stamp, ( ) and stuck the stamp on the letter. ( ) He gave the stamp a good bang with his fist ( ) so it couldn't possibly come off. Then he went round the corner to the letter box, and pushed his letter in. ( ) Then, crossing his fingers for luck, ( ) he went home. His mum gave him an ice lolly to lick ( ) that she had kept in the freezer. It seemed a very long time to wait for the whole week till the results of the competition came through, but the end of the week came at last. Ted heard the postman knock on the door ( ) and rushed to open it ( ). 'Something for you' said the postman, and handed Ted a huge envelope. ( )

Ted tore it open with shaking fingers. ( ) He couldn't read all the writing on the page inside, but at the bottom was a picture of a computer game so he guessed it was good news. He ran to give it to his mum. She read it carefully, ( ) and then gave Ted a big smile. ( ) 'You did it!', she said proudly. Third Prize! We can go and choose your computer game this afternoon.' When they went, they saw Ted's picture and some others pinned up on a big display in the shop. That day, Ted was in a very, very GOOD mood!

Who is Asleep?

Level I

Awareness of Others

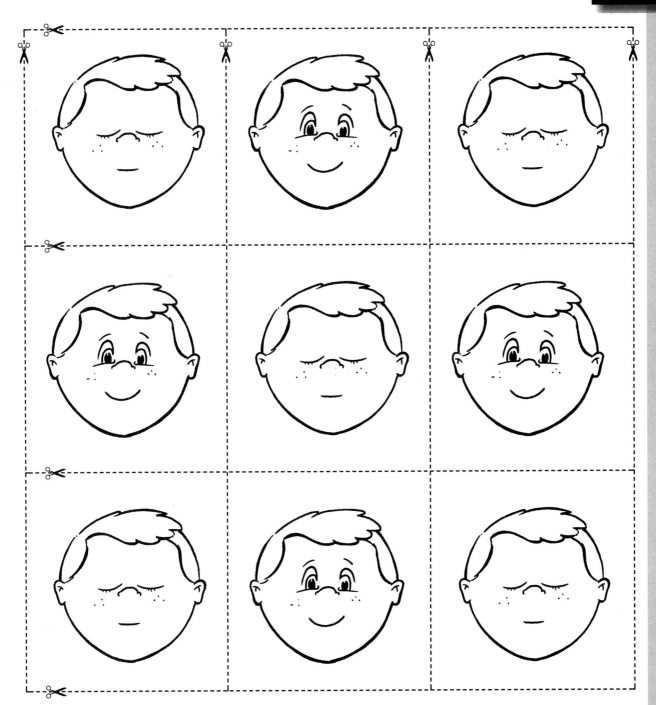

PERSONAL AND SOCIAL DEVELOPMENT RESOURCES

Lions and Tigers

Level III

Confidence & Independence

Lions and Tigers *(continued)*

Level III

Confidence & Independence

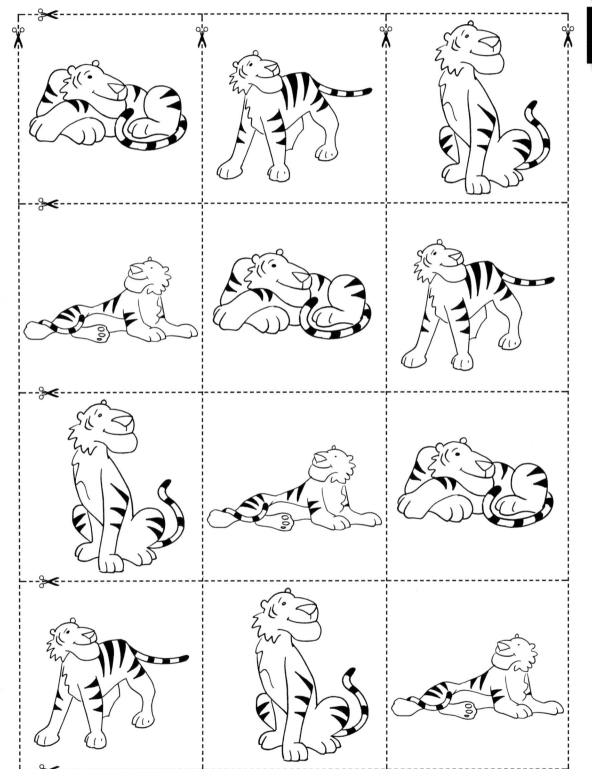

Scavenge Hunt

Level III

Confidence & Independence

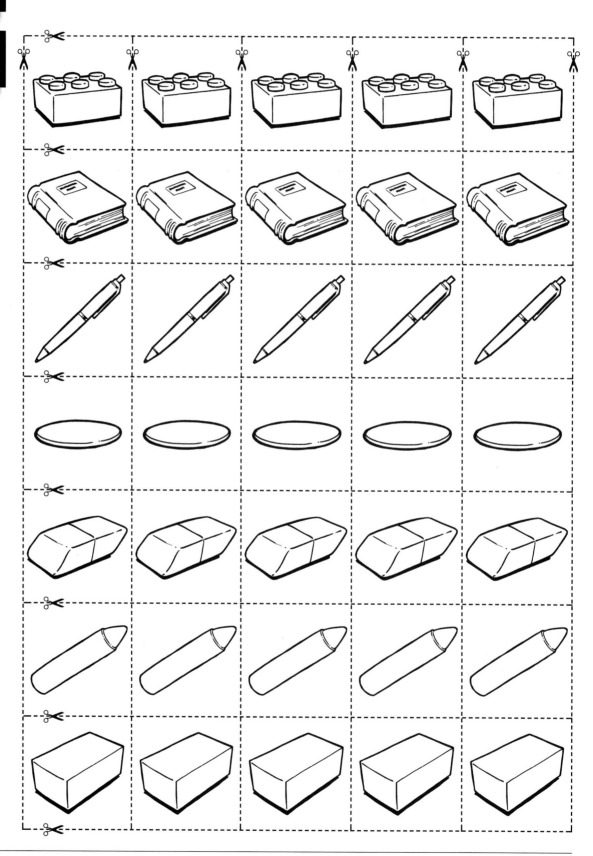

RESOURCES   PERSONAL AND SOCIAL DEVELOPMENT

Happy–Sad

Level I

Feelings & Emotions

DEVELOPING BASELINE COMMUNICATION SKILLS   PAGE 237

Happy–Sad *(continued)*

Happy–Sad *(continued)*

happy

Happy–Sad (continued)

sad

RESOURCES   PERSONAL AND SOCIAL DEVELOPMENT

Angry–Scared

Level II

Feelings & Emotions

DEVELOPING BASELINE COMMUNICATION SKILLS   PAGE 241

Angry–Scared (continued)

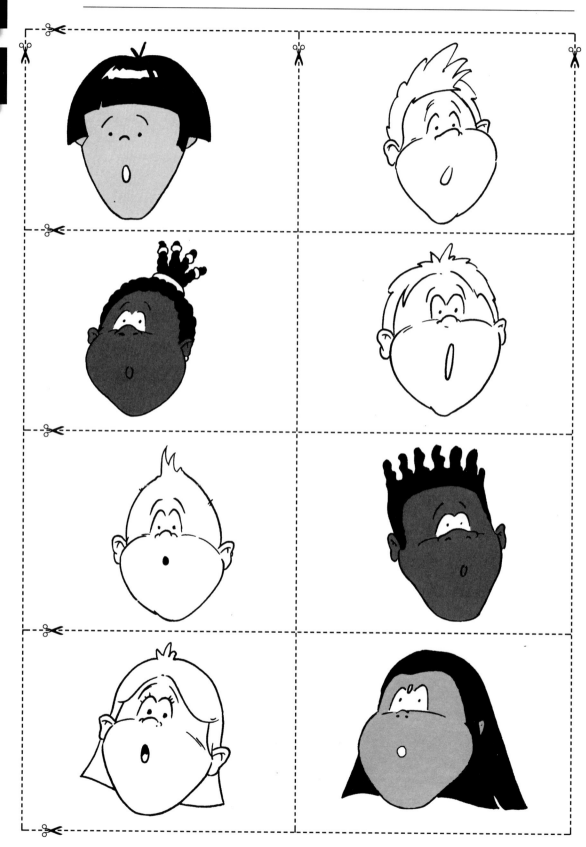

Angry–Scared (continued)

angry

Angry–Scared (continued)

scared

All Change

Level II

Feelings & Emotions

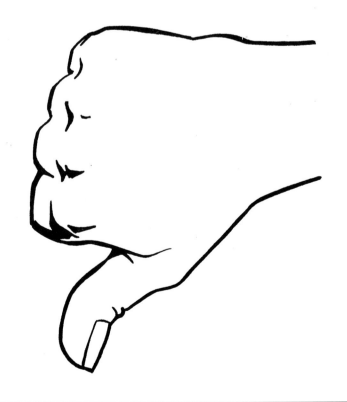

Spin-a-Word

Level IV — Feelings & Emotions

1 Boring, boring

Toby got bored very easily. Whatever his mother or father suggested he might like to do, Toby always answered (*'boring'*). Saturdays were the worst, because there was no school. One Saturday, it was pouring with rain and Toby's best friend Darren was away staying with his grandmother. 'How about some colouring?' asked Toby's mother. Toby said (*'boring'*). 'Well, what about a video?' Toby said (*'boring'*), 'I've seen all my videos hundreds of times.' 'Help me make a cake?' Toby said (*'boring'*). 'Help dad in the workshop?' Toby said (*'boring'*). 'Well,' said his mum, getting desperate, 'You could make that new Airfix model you got for your birthday.' Toby was really cross now. At the top of his voice he shouted 'Very very VERY (*'boring'*).'

Mum was getting pretty cross too. 'How about inviting a hundred and two pink elephants to tea?', she said. Toby said (*'boring'*), and then 'WHAT did you say?' Toby had a bit of a giggle then. 'Now you're in a better mood', mum said 'Would you like to come with me to see the *Lion King*?' Even Toby couldn't say that would be (*'boring'*).

(Substitute the latest in-film for the *Lion King*!)

Spin-a-Word *(continued)*

2 Exciting

James and his sister Rebecca had gone to the fair. Most of all they wanted to go on the Giant Switchback. 'Please, dad', said James, 'It would be so (*'exciting'*). 'Lots of the other things are (*'exciting'*)' said dad. 'Are you sure you want to go on the Switchback? It might be so (*'exciting'*) that its almost scarey.' 'We like things to be (*'exciting'*) said James and Rebecca together. 'All right', said dad, 'Here's the money. I think I would find it too (*'exciting'*). I'll just watch from down here. Don't forget to hold on tight.'

James and Rebecca paid the man, and climbed into one of the little cabins. There was a grumbling engine sound, and they began to move off, slowly at first and getting faster and faster. Rebecca shouted 'This is really (*'exciting'*)'. 'Faster, faster' shouted James, 'I want it to get even more (*'exciting'*)'. They whirled up and down and round corners, faster and faster, and if they looked down they could just see dad looking up and waving to them. When they got down their legs were wobbly and their heads were spinning. 'I think I feel a little bit sick', said Rebecca, 'But that was really so (*'exciting'*.)'

'I hope that will keep you quiet for a little while', said dad.

Spin-a-Word (continued)

3 Surprising

It was April Fool's Day. That's the day when people play tricks on each other, and try to catch each other out. Sam and his sister Amy had been up for ages, arranging little tricks around the house. When mum got up and went downstairs, they held their breath waiting for her to discover surprises they had planned for her. Mum was still half asleep. She went to the fridge, opened the door, and reached for the milk. There was a big fluffy Furbie sitting in the fridge. [Substitute the latest in-toy.] 'This is rather (*'surprising'*), 'said mum, 'How did that get in there?' She made her tea, put in a teaspoonful of sugar, and took a big sip. 'Ugh!' said mum, making a face. 'That was (*'surprising'*) I seem to have put in salt instead of sugar. How did I come to do such a silly thing?' Sam and Amy tried hard not to giggle. They had put salt in the sugar bowl. Dad came down just then. He always liked a boiled egg for breakfast, and Sam and Amy had found a pretend one in the joke shop. When dad wasn't looking they popped it in his eggcup. Dad gave it a bang with his spoon. Nothing happened. He banged again. 'This is rather (*'surprising'*)' said dad, 'I don't seem to be able to break this egg.' He picked it up and took a look. Then he realised, and turned round so he could see the calendar. Sam and Amy made signs to him not to tell mum. At that moment they saw the postman coming past the window. They rushed to the door, and quietly took in all the post. 'What is there for me?' called mum. 'Nothing' said Sam. 'That's a bit (*'surprising'*)' said mum, 'I was expecting a parcel.' 'No, nothing here' said Amy. 'Well I do call that (*'surprising'*)' said mum again. She sounded so disappointed, Amy and Sam couldn't keep it up any more.

'April Fool!' they shouted. 'Well I'm bothered', said mum, laughing, 'I don't usually get taken in like that. How VERY (*'surprising'*)!'

Spin-a-Word *(continued)*

4 Cross

When Jack woke up in the morning, he knew it was going to be a bad day. He felt really (*'cross'*). He didn't know why, he just felt (*'cross'*). And of course when you feel (*'cross'*) things always go wrong, and you feel worse and worse. So when Jack got out of bed, he fell over the train track he'd left on the floor, and banged his knee. He stopped to play with his train for a bit, and got late, and mum started calling up the stairs that if he didn't hurry up, he'd miss the school bus. So then in his hurry he put both legs down one side of his trousers and got in a terrible muddle. When he got downstairs mum took one look at his face and said 'Jack, you DO look (*'cross'*). What's the matter with you? Did you get out of bed on the wrong side?' 'Don't know', Jack muttered, 'I just feel (*'cross'*)'. 'Well, have some cereal', said mum, 'But we've finished your favourite I'm afraid. That will make you (*'cross'*) too, I expect.'

When Jack had finished his breakfast, and was slamming about trying to find his coat and his school bag, mum came out of the kitchen to find him. 'Shall I tell you what I do when I start the day (*'cross'*)?' she said, 'I go back upstairs and pretend it's a new day, and start all over again. And it's your pocket money day, so you can take this up with you.' So Jack went upstairs, lay down on his bed, got up again, and went downstairs all over again. It was true, he didn't feel nearly so (*'cross'*) now, and he had his pocket money to spend. Perhaps it would be a good day after all.

Level I

Feelings & Emotions

Spin-a-Word *(continued)*

5 Frightened

It was Bonfire Night. Joshua and Emma lived in a village which had a village green and a pond in the middle, and an enormous bonfire had been built on the green. On top of it was a big Guy Fawkes with a head made out of a turnip, and wearing a floppy hat. There was going to be a party after dark, with sausages and baked potatoes to eat, and fireworks. 'We'll need to keep the pets in', said mum. 'When the fireworks go off they may be (*'frightened'*) and try to run away, so they need to be safe in the house.' Joshua had a puppy called Barley, and Emma had a cat called Twitch.

It did seem a long day waiting for the party and the fireworks to begin. They had to keep remembering to shut doors behind them, and they couldn't open the windows because of Twitch. At last six o'clock came. 'Animals both safely in?' asked mum. 'We'll watch the first fireworks from here, and make sure they are all right.' Just then the first rocket went off with a zoom and a whizz and a bang. Twitch jumped down from the table and crept under an armchair. 'She is (*'frightened'*)' said mum. 'That was a noisy one.' Then off went another rocket, and another, and a squib. That was too much for Barley. He rushed upstairs and hid under a bed. 'He's very (*'frightened'*)', said Joshua. He went upstairs after Barley, and tried to coax him out, but Barley was too (*'frightened'*) to leave the safety of his dark hiding place. 'Why are they (*'frightened'*)?' Emma asked mum. 'I don't know' mum said. 'I think it's just that they don't know what those sudden loud noises are. I can remember when I was little I had a dog that was very (*'frightened'*) by fireworks, and by thunderstorms. I don't think he knew which was which! But we'd better go out now, or we'll miss the rest of the fireworks, *and* the sausages and baked potatoes.' By the time the family came indoors again, Barley and Twitch had come out of hiding, and were in the kitchen finishing up their dinners. 'No need to be (*'frightened'*) any more', said Emma. 'It's all over till next year.'

Spin a Word *(continued)*

Level IV

Feelings & Emotions

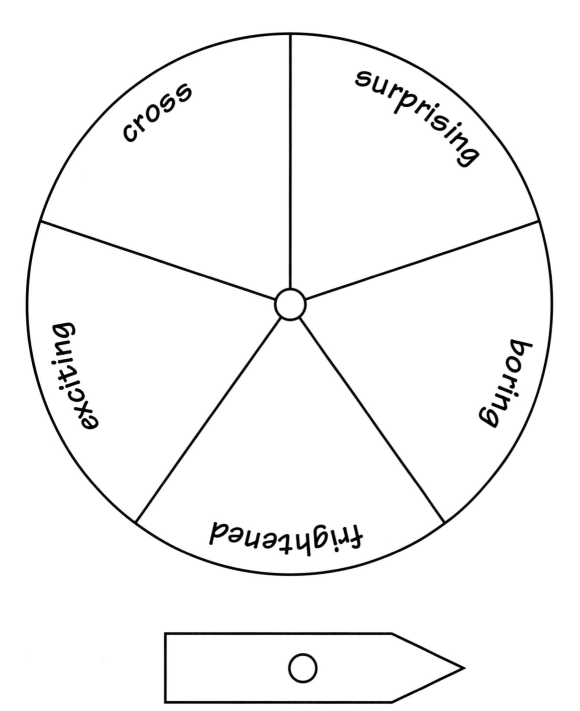

Where's Granny Going?

Further scenarios:

Granny has:
A long pole with string on, a stool, a box of maggots, a pail.

Walking boots, an anorak, a map, a walking stick.

Her cat in a basket, her purse.

A spade, a fork, a trowel, a wheelbarrow, some plants.

A duster, a mop, the hoover.

A ladder, a bucket of water, some cloths.

A big roll of paper, a bucket of glue, a pasting brush, a ladder.

Airline tickets, travel pills, a hotel brochure.

A basket, boxes of sandwiches, cakes and biscuits, a rug, a flask of juice.

A bag full of books, a reading list.

A suitcase, lots of books to read, her medicine bottles.

A present wrapped in pretty paper, an invitation card, a funny hat, a balloon.

It's a Funny World

Level IV

Understanding

To be read *slowly* to allow children time to spot the absurdities, which appear in italics. Children's possible responses are in parentheses.

1 Lee and Daniel set off together down the street. They passed a *fish walking* the other way (Fish can't walk. They haven't got any legs/feet.) The *fish smiled* at them and *said 'Good Morning!'* (Fish can't smile and can't talk.)

The pavements were very crowded, and Lee and Daniel were worried the toy shop would be shut before they got there, so *they flew* over some houses and into the next street. (Boys can't fly. They haven't got wings.)

They got to the toyshop just as *an elephant was shutting the door* and locking up. (Elephants don't lock up shops.) 'Oh please ' said Lee and Daniel, 'Please let us in. We've got some pocket money to spend and we want to buy some Lego®.' *'Oh all right then' said the elephant.* (Elephants can't talk.) So Lee and Daniel went into the toy shop and *swam* to the Lego® stand. (You can't swim in a shop.) They chose some new Lego® with wheels and pulleys so they would be able to build a windmill. Then they took the box to the check out. Lee looked in his purse and *took out some oranges* to pay the lady. (You don't pay with oranges, you pay with money.) 'Thank you' said the lady, and gave them some change. Lee and Daniel ran home. Their mum *gave them some worms for tea* (you don't eat worms), and after tea they built a huge windmill with the new Lego®.

It's a Funny World *(continued)*

2 Dad was going fishing. He packed some sandwiches for his picnic, and a big flask of *toothpaste*. (You don't drink toothpaste.) In case it rained he took *a sun hat* (you don't wear a sun hat in the rain), and he also took a box of wriggly maggots for bait. Then off he went to the river *in his space rocket* (dads don't have space rockets). When he got to the river bank, dad sat down and got out his fishing rod. He stuck an *old boot* on the hook, (you don't catch fish with an old boot, you catch fish with maggots), and dangled it in the water. Nothing happened for ages. Then all of a sudden a fish *put an arm up* out of the water (fish don't have arms) and grabbed hold of the hook. Dad pulled the fish in and put it in his bucket. Then he thought he would have his picnic. He got out his nice *grass sandwiches* (you don't eat grass) and *ate them with his nose*. (You don't eat with your nose.) Then *he drank his tea with his ear*. (You don't drink with your ear.) Then dad lay down and had a little sleep *in the river*. (You don't go to sleep in water.) When he woke up it was getting quite late and it was beginning to rain. Dad *put on his sun hat* (you don't wear a sun hat in the rain) and went slowly home.

It's a Funny World *(continued)*

3 Jamie and Oliver knew it was time for school. *It was the middle of the night*. (You don't go to school in the middle of the night.) They got dressed in their best *school pyjamas* (you don't wear pyjamas to school) and brushed their teeth *with the hoover*. (You don't brush your teeth with a hoover.) Then they collected their lunch boxes and went to wait for *the school helicopter*. (You don't go to school in a helicopter.) On the way to school they had a look in their lunch boxes. Jamie had *plasticine biscuits* (you can't eat plasticine) and some *petrol to drink*. (You don't drink petrol.) Oliver had *octopus sandwiches* (you don't eat octopus sandwiches) and a nice drink of *ink*. (You don't drink ink.) When they got to school they hung their coats up *on a tree* (coats are hung on pegs not trees) and went into the classroom. Their teacher was not very well that day, so *a hippopotamus was taking the lesson*. (A hippopotamus can't teach.) Jamie and Oliver sat down *on the ceiling*. (You don't sit on the ceiling.) At playtime, all the children went to play football *on the roof*. (You can't play football on the roof.) After play, Jamie did some very good writing *with a banana* (you can't write with a banana) and Oliver did some painting *with a brick*. (You can't paint with a brick.) At the end of the day, both boys got smiley faces *for doing very bad work*. (You get smiley faces for good work.) They were really tired after working so hard.

Level IV

Understanding

Level IV

Understanding

Work it Out!

1. When he heard the car door bang he ran to the door, wagging his tail and barking loudly.

2. He got out of his red van and carried a pile of letters and a parcel up the path. The parcel wouldn't fit through the letter box so he rang the door bell.

3. She woke up early and put on the new school uniform. The sweatshirt was blue and the skirt was grey, and she had a new pair of white socks too. She felt excited about her first day at school, and a bit nervous too.

4. Mum and dad were smiling at him, and on the table there were some presents and a pile of cards. He opened the first card. It had a picture of a boy on a bike, and a badge with a big red '7' on it.

5. She crouched under the bush, watching the mouse coming out of a hole in the hedge. Her striped tail flicked from side to side. Then she pounced … but the mouse was too quick for her.

6. It was a cold windy morning. She stood by the side of the road, in her bright yellow coat and hat, holding the 'STOP!' sign. Then a group of mothers and children appeared. The children smiled and said hello, as she picked up the sign and stepped into the road to stop the traffic and allow them to cross the road.

7. She walked slowly, using a walking stick to help her. Every few minutes she had to stop and have a rest. It was a windy day, and her white hair blew across her face. When she got to her house she fumbled in her basket to try to find the key. Her fingers were bent and it was hard to pick the key up.

Work it Out! (continued)

Level IV

Understanding

8 He crawled over to the table and reached up with his chubby little hands. His fingers grabbed the edge of the table and he managed to pull himself up. Now he was standing, wobbling a bit and holding on to the edge of the table. He could just see what was on the table – a shiny toy train. He wanted it, so he reached forward with one hand, and suddenly sat down hard, with a bump.

9 It was great swinging like this. She discovered that if she moved her legs right, she could make the swing go higher and higher. It was like flying. When the swing got right up to the top she could see her little brother looking up at her. He was too small to swing this high. When the swing came down she could feel the wind rushing at her face. The next time the swing came down she jumped off, and ran across the park to join her friends.

10 He walked into the classroom and put the pile of books on his desk. It was raining outside. Soon the door opened and 30 boys and girls crowded into the room, chattering and pushing each other. He stood up, and said, loudly, 'Good morning everyone'. The chattering stopped.

11 He could see the road stretching in front of him. He could hear the thud, thud of his trainers on the road, and the sound of his breathing, loud in his chest. His legs were muddy where the puddles had splashed them. At last he saw his house, at the bottom of the hill and he knew he would soon be home.

12 The castle was nearly finished now. He filled one more bucket with sand and carefully tipped it onto the top. Then he stuck one of the little flags his dad had bought on the very top. He stood up and looked around. His sister was sunbathing on a rock. He called her over to look at his castle.

13 She put on her white coat and washed her hands. Then she looked in her diary. The first person on the list was Mrs Jones. She opened the door and looked into the waiting room. Mrs Jones was sitting by the window, and Tiddles the cat was crouching in his cat basket, on the floor beside her.

LANGUAGE AND LITERACY   RESOURCES

Big Green Apples and Only One Left

Level I

Speaking

These pictures have to be coloured before use.

PAGE 258   DEVELOPING BASELINE COMMUNICATION SKILLS

Colour Families

These pictures have to be coloured before use.

Level II

Speaking

George the Giant

RESOURCES　　　　　　　　　　　　　　　　　LANGUAGE AND LITERACY

George the Giant *(continued)*

Level II

Speaking

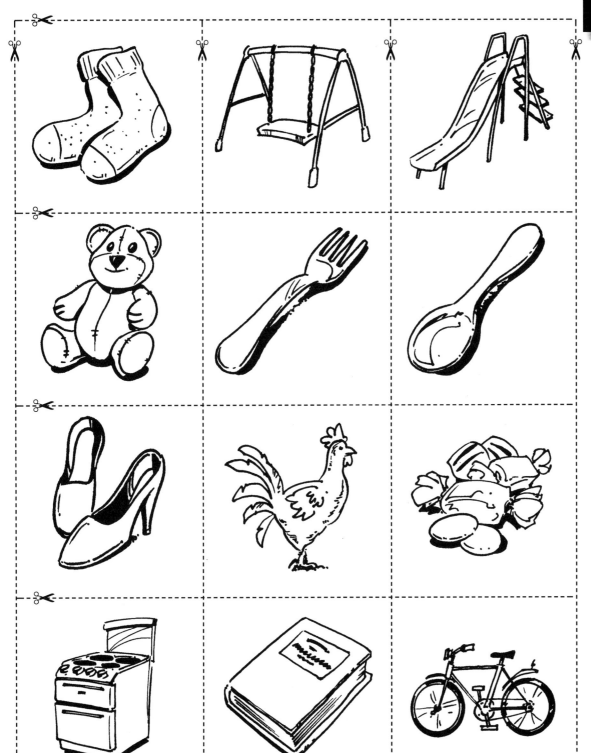

DEVELOPING BASELINE COMMUNICATION SKILLS

Level II

Speaking

George the Giant *(continued)*

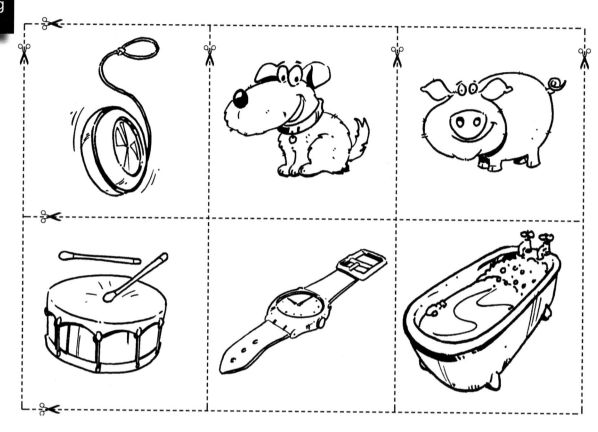

Spot the Difference

Spot the Difference *(continued)*

Zig at the Zoo

STORY

It was Saturday, and Zig decided to go to the zoo. He took his lunch in his Batman lunch box, and put it in a plastic bag. It was a hot day, and he bought himself an ice cream – a strawberry one with a chocolate flake on the top. First he visited the monkeys. He loved watching them doing all sorts of silly tricks. There were four monkeys playing around. They snatched fruit from each other, swung by their tails from the pretend trees, and ran round and round the cage. Zig was so busy laughing at the monkeys that he didn't notice a little brown and white one, with long skinny arms. It was clinging to the side of the cage and its round brown eyes were fixed on the ice cream cone.

As it was such a hot day, Zig put on his bright green baseball cap. He had bought it when he went to Disney World, and he thought it looked really cool. He strolled over to the penguins, who were keeping cool in their deep green pool, while they waited for the keeper to come along with his bucket of fresh fish. He leant over the side of the fence, looking at his reflection in the water. A penguin dived off a rock just beside where Zig was standing, and he leaned a bit further forward, to try to catch sight of it under the water.

Near the penguins was another, bigger pool, where the seals lived. Zig loved watching them doing tricks. They could balance balls on their noses, fetch plastic ducks from the water, and catch coloured rings that the keeper threw for them. There were chairs arranged in rows, so that people could watch the show, which was at three o'clock every afternoon. It was nearly three o'clock now, so Zig took a seat in the front row. Soon the keeper arrived, carrying the balls, ducks and coloured rings in a big red bucket. After he had thrown a few balls, which the seals retrieved and balanced on their noses, showing off to the audience, he invited members of the audience to come and help him. Zig stood up, and waved his paw at the keeper. 'You, in the front row, the one with the green baseball cap, come on up here and help me,' said the keeper. Zig was very excited. He climbed up on the rock beside the keeper. 'Right

Zig at the Zoo (continued)

then,' said the keeper, 'take one of the green coloured rings and throw it as far as you can, across the pool. Mind you don't overbalance!' Zig took the ring, raised his paw as high as he could and …

Zig was feeling hungry. He walked over to the 'Pets' Corner'. Here there were lots of tame animals, running free, and you were allowed to touch them. There were benches to sit on, and Zig decided this would be a good place to have lunch. He sat on a bench in the shade of a large tree and opened his Batman lunch box. Inside was a cheese and tomato roll, a carton of orange juice, a packet of crisps and his favourite, a beautiful, ripe peach. He would save the peach until last! He ate the roll first, and drank the juice. Then he opened the crisp packet, and put the lunch box under the bench to keep cool. He sat back, munching his crisps and watching the goats, the rabbits with long floppy ears and the lemurs. The lemurs were about the size of a large cat. They had huge eyes, grey fur and long black-and-white striped tails. They were very tame, and one of them came right up to where Zig was sitting. He reached under the bench to get his peach. Oh dear!

After lunch Zig decided to go and visit the camels. There were five of them, in a large field. They had rather scruffy brown coats, and looked very proud and rather bad tempered. Zig knew that in some parts of the world people actually rode on camels, but he didn't like the idea himself. The camels stood in a group and every few moments they made loud grunting sounds. Zig had also heard that camels can spit, but he didn't know if this was true or not. After a while one of them walked slowly over to where Zig was standing. Zig looked up at the camel and the camel looked down at Zig. Slowly it lowered its head, until it was almost level with Zig. It half closed its eyes and all of a sudden . . .

Zig was getting quite tired now. He would go and visit the Tropical House, then it would be time to go home. The Tropical House was hot and steamy. Brightly coloured butterflies darted in and out of the lush

Zig at the Zoo *(continued)*

green trees. If you looked carefully you sometimes caught sight of the bright blue wings of a humming bird as it took nectar from enormous red and pink flowers. In the pond goldfish swam lazily around, and Zig heard frogs croaking under the shade of the large lily leaves. Suddenly Zig heard a voice, which seemed to come from the tree he was standing by. 'Mind out, the lions are out.' Zig looked up, but he couldn't see anyone. Then he heard it again, 'Mind out, the lions are out.' He began to feel a bit nervous. How was he going to get back to the Way Out if there were lions prowling around. He heard the voice again, and this time it carried on talking, 'Pretty Polly, wipe your feet'.

LANGUAGE AND LITERACY　　　　　　　　　　　　　　　　RESOURCES

Level IV

Auditory Memory

Pantomimes

Text 1

This game will work out best if you make one group into the monkeys and one group into the elephants. Point to the groups in turn asking 'What did he say?' or 'What did they say?' and wait for the children to tell you.

Once they are in the swing of it, you will probably not need to say 'What did he say?' but just to point to the appropriate group. The pauses for asking or pointing are indicated by ( ).

One day, in the jungle, the monkeys were teasing the elephants. They were hiding and then jumping out at them; they were whizzing up trees and throwing fruit and twigs at them; they were shouting cheeky things at them; and the elephants were getting crosser and crosser. One little monkey in particular was being really naughty. He shouted out 'Can't catch me' (point to the monkey group and ask 'What did he say?'). The elephants trumpeted 'Oh yes we can' (point to the elephant group and ask 'What did they say?'). 'Oh no you can't' (point to monkeys). 'Oh yes we can' (point to elephants). The monkey squeaked 'I'm up here'.( ) The elephants bellowed 'We're coming to get you'. ( ) The monkey teased them 'You can't climb trees'. ( ) 'Oh yes we can'. ( ) 'Oh no you can't'. ( ) The elephants roared 'You just watch'. ( )

And the elephants started to try to climb the tree. They slipped and they slithered, and all the time the little monkey was throwing down twigs and leaves and bananas, and laughing and giggling to himself. 'Come a bit higher', he giggled. ( ) 'You come down here!' roared the angry elephants. ( ) Suddenly one of the elephants had an idea. They all wrapped their trunks round the monkey's tree, and began to pull it – this way and that, this way and that. Up in the top of the tree the monkey was swaying about, and it seemed that at any moment the tree would come crashing down. The little monkey was scared. 'Stop rocking my tree,' he squeaked. ( ) 'Only if you say please,' growled the elephants. ( ) 'Please stop rocking my tree.' ( ) 'Say it once more then,' said the elephants. ( ) 'Please stop rocking my tree.' So then the elephants unwrapped their trunks, gave a last trumpet at the little monkey, and went off to sleep in the sun. The little monkey whispered 'I'll tease you again tomorrow.' ( ) But he didn't dare say it out loud.

Pantomimes *(continued)*

Text 2

Do you remember the story about the naughty monkey and the elephants? Well, about a week later the monkey was feeling very naughty again. All the elephants were sleeping in the sun near a big pond. The naughty monkey crept up near them and hid behind a bush. Then he made a noise like a tiger. Elephants are quite scared of tigers, so they woke up in a fright. 'What was that?' asked the elephants. ( ) Using his biggest voice, the monkey shouted 'It's a tiger.' ( ) 'Oh no, a tiger.' said the elephants. ( ) 'Two tigers,' shouted the monkey. ( ) 'Oh no, two tigers,' said the elephants, shaking with fright. ( ) 'It's three tigers,' shouted the monkey. ( ) 'Oh no, not three tigers,' said the elephants, really scared. ( ) But just then one of the elephants saw the monkey's naughty face peeping out at him from behind a bush. 'Its that pesky monkey,' trumpeted the elephant. ( ) 'It wasn't me,' squeaked the monkey. ( ) 'Oh yes it was,' bellowed the elephants.( ) 'No it wasn't really'. ( ) 'Oh yes it was.' ( ) 'And we're coming to get you.' ( ) Then the elephants made a big circle round the naughty little monkey, and they filled their trunks with water, and they squirted him all over as hard as they could. 'Oh stop it, stop it,' squeaked the monkey.( ) 'Say you're sorry then,' bellowed the elephants. ( ) 'I'm sorry, I'm sorry,' squeaked the monkey. ( ) 'All right then, we'll stop,' said the elephants. ( )

'Don't do it again.' ( ) 'Oh yes I will,' whispered the monkey. ( ) But he didn't dare say it out loud.

Level IV

Auditory Memory

Pantomimes (continued)

Text 3

Shadow the Sheepdog was out in the field, trying to move a flock of sheep through a gate in the corner. But the sheep were in an awkward mood, and they wanted to stay where they were. Shadow raced backwards and forwards, barking and barking, and the sheep would move a little bit, and then stop and begin eating grass again. 'Oh do get going,' barked Shadow. ( ) 'We like it here,' baaed the sheep. ( ) 'I can't help that,' barked Shadow. ( ) 'This grass is better,' said the sheep. ( ) 'I'll fetch the farmer,' barked Shadow. ( )'Go on then, silly,' said the sheep. ( ) 'I'll give you a nip,' barked Shadow. 'Oh no you won't,' said the sheep. ( ) 'Oh yes I will.' ( ) 'Oh no you won't,' said the sheep. ( ) 'We'll go in a minute,' said the sheep. ( ) 'You've got to go now.' ( ) Just then the farmer came out in his tractor, and brought the other two sheepdogs with him. 'I see you're having trouble,' said the farmer, and he began to drive the tractor behind the sheep, while the three sheepdogs ran here and there like lightning, barking and barking. 'All right, all right, we're going,' muttered the sheep, ( ) and they began to move obediently towards the gate. Through they went, one, two, three, one after the other, until they were all in the next field, and the farmer slammed the gate.'I'll help you next time,' he said to Shadow.

Fact Finder

Level IV

Auditory Memory

MY FAMILY

My name is Kyle and I'm seven years old. I have two sisters who are older than me, and a baby brother. My dad works at the bank. I wish I could have a puppy, but mum says our house isn't big enough. We have got a hamster though – he's brown and white and he's called Scruffy, and you have to be careful because if you put your fingers through the bars he bites.

How many children are in the family?
Where does their father work?
What is the boy's name?
How old is he?

AT THE CASTLE

The castle is on top of a hill. The king and queen live there with Princess Zoe. So does a wicked witch. She lives in the dungeons with her pet rat, Gus. Gus is really a prince, but the wicked witch made a spell and turned him into a rat. He hates living in the dungeons with the witch, but he cannot change back into a prince until Princess Zoe is 16 years old.

Where is the castle?
What is the princess' name?
Who lives in the dungeons?
What is the rat called?
What happened to the prince?

THE PICNIC

My friend came over on Saturday and mum let us make a picnic. We made jam sandwiches. Then we cut up some pieces of cheese. We put some raisins and some chocolate chips in a pot, and we found two bars of chocolate in the cupboard. Mum gave us a carton of juice each. We put everything in a bag and went into the garden. There is a hedge at the end of the garden, with a gap in it, just big enough to sit inside. So we got inside, and ate our picnic and we pretended we were explorers in the jungle. Suddenly there was a rustling noise in the hedge. What could it be? A black furry shape appeared. It was our cat, Smokey, and he wanted to eat our crisps.

Level IV

Auditory Memory

Fact Finder *(continued)*

What kind of sandwiches did they make?
What else did they have for their picnic?
Where did they eat the picnic?
Where were they pretending to be?
What was the cat's name?

THE CLOWN

Juggling Jim is a clown. He lives in a bright yellow caravan with Screecher, his parrot. Jim wears baggy trousers with one red leg and one blue leg. His shirt is yellow with pink spots, and when he goes outside he always wears a yellow bobble hat. I don't know why he's called 'Juggling Jim', because he's no good at juggling. He always drops the balls. What he *is* good at is making people laugh, especially children. He can do lots of funny things; he can waggle his ears; he can touch his nose with his tongue; he can balance an egg on the top of his head. But the funniest thing he does is eat a bowl of jelly while he rides his bike.

What is the clown's name?
What colour is his caravan?
What does he wear on his head when he goes outside?
Name three funny things he can do.
What is the funniest thing he does?

A DAY AT THE BEACH

Joe looked out of his bedroom window. It was a hot sunny day, and he was really excited because his mum was taking him to the beach for a whole day. His mum said his friend Sam could come too. Joe was six-years-old and Sam was five. After breakfast they got in the car and went to collect Sam. Then they set off. Sam and Joe sat in the back singing. Every few minutes Joe said 'Mum, are we nearly there?' Suddenly there was a bang from the engine, and the car stopped. Steam was coming out from under the bonnet. Sam and Joe looked at each other. Mum got out and looked worried. She told the boys to get out of the car, and they all stood by the side of the road, wondering what to do. Joe began to feel sad. Now they wouldn't be able to go to the beach. He wanted to kick the car. Just then another car stopped. It was mum's best friend, Sally. Mum explained about the engine, and Sally said 'Let the boys come with us – we're going to the beach too.'

Fact Finder *(continued)*

What were the two boy's names?
How old was Joe?
What was the weather like?
Where were they going?
What went wrong?
Who helped them?

SHOPPING TRIP
On Saturday Kate and her auntie Sally went shopping. They caught the bus into town. First they went to the shoe shop and auntie Sally spent ages trying on different pairs of shoes. Then they went to the book shop, and auntie Sally looked at lots of different books. It seemed as if she was reading them all. After that they went into a clothes shop. Auntie Sally wanted a new dress, and she kept taking different ones into the changing room. She didn't seem to like any of them. Kate felt very bored. At last auntie Sally bought a dress, and then said 'Let's go to a café'. Kate chose a milkshake and a cake. Auntie Sally had some coffee. Then she reached in her bag, and said, 'Look what I've got for you.' There was a pair of pink fluffy slippers from the shoe shop. There was a book about ponies from the book shop, and last of all, there was a bright pink sweatshirt and a pair of stripey socks from the clothes shop!

Who went shopping?
How many shops did they go to?
What was Kate's book about?
What did auntie Sally have to drink?
What did auntie Sally buy?

Jack and Jill

Jack and Jill (continued)

INSTRUCTIONS

    Put a red dot on Jack's hat

    Put a blue cross on Jill's nose

    Put a green dot on Jill's hat

    Put a yellow circle on Jack's hand

    Put a purple circle on Jill's hand

    Put a red cross on Jill's skirt

    Put a blue circle on Jack's trousers

    Put a green line on Jack's shirt

    Put a green line on Jill's jumper

    Put red dots all over Jack's shoes

    Put pink dots all over Jill's shoes

Level IV

Auditory Memory

Pitter Patter

After the first game, you can leave the groups as they were but give them different words to say:

Raindrops	– Splishy sploshy
Rabbits	– Skimper scamper
Hares	– Flippety floppety
Kangaroos	– Gallopy ballopy

or exchange the raindrops with the rabbits, and the hares with the kangaroos.

After this you may want to change the groups:

Witches	– Abracadabra
Wizards	– Wizzly wozzly
Goblins	– Migicky magicky
Trolls	– Finicky fanicky

And all change again!

Butterflies	– Flittery fluttery
Birds	– Twittery wittery
Worms	– Slithery swithery
Moles	– Grumbly bumbly

On the Beat and Prove It

SYLLABLE WORD LISTS

Single Syllable

cheese	jam	clock	jug
bath	fish	book	key
watch	peach	pig	snail
pen	tree	chair	square
hat	car	snake	boy
hill	bear	bee	leaf
fence	shell	sun	lake
hand	church	bowl	cliff
six	tooth	cup	nail
bus	shirt	light	sock

Two Syllables

rabbit	balloon	pencil	sandwich
finger	twenty	water	chimney
paper	ruler	orange	garden
rubber	picture	glasses	kitten
carpet	castle	cupboard	ladder
table	window	butter	baby
apple	beetle	mirror	zebra
jumper	river	button	sausage
lorry	flower	cooker	parrot
money	burger	scissors	lolly

Level III

Level IV

Phonological Awareness

On the Beat and Prove It

Three Syllables

radio	dinosaur	rectangle	submarine
aeroplane	crocodile	caravan	spaghetti
triangle	octopus	butterfly	gorilla
piano	magazine	buttercup	India
violin	cucumber	daffodil	Africa
computer	tomato	motorway	pyramid
video	potato	animal	grasshopper
telephone	microwave	unicorn	calendar
bicycle	holiday	cinema	strawberry
cereal	microphone	telescope	

Four Syllables

helicopter	alligator	caterpillar	binoculars
television	radiator	America	brontosaurus
supermarket	calculator	thermometer	stegosaurus

TV Tongue Twisters

It is important to explain the *meaning* of each tongue twister to the children before they attempt to say it, particularly where there is an obscure word such as 'flagon'.

Examples:

Sister Susie's sewing socks for sailors

Round the ragged rocks the ragged rascals ran

Which switch is the switch for Ipswich?

Use the stairs with the squares on the stairs

See the butterfly flutter by

See the dragonfly drink a flagon dry

Sleepy snakes only slide slowly

There's a moth in my scotch broth

Who put slippery kippers in my slippers?

The train trundles down the twisty track

Hugh drew two new blue shoes

The cat crept closer and closer to the cream

Red lorries, yellow lorries, red lorries, yellow lorries, red lorries, yellow lorries

Level IV

Phonological Awareness

Developing Baseline Communication Skills
CROSS-REFERENCE TABLES

Personal and Social Development

282 / Turn-Taking
283 / Body Language
284 / Awareness of Others
285 / Confidence and Independence
286 / Feelings and Emotions

Language and Literacy

287 / Understanding
288 / Listening and Attention
289 / Speaking
290 / Auditory Memory
291 / Phonological Awareness

TURN-TAKING

Level	Activity	Phonological Awareness	Auditory Memory	Speaking	Listening and Attention	Understanding	Feelings and Emotions	Confidence and Independence	Awareness of Others	Body Language
LEVEL I	Musical Hat				■					
LEVEL I	Talking Toy (i)			■				■		
LEVEL I	Group Lotto									
LEVEL I	Balloon Bubbles (i)									
LEVEL I	The Farmer Wants a Horse							■	■	
LEVEL II	Talking Toy (ii)			■						
LEVEL II	Pull Out a Name									
LEVEL II	Number's Up!				■					
LEVEL II	Add to It				■					
LEVEL II	Nursery Rhyme Circle			■				■		
LEVEL III	Feely Bag Game			■				■		
LEVEL III	Yes–No									
LEVEL III	In My Case		■	■	■					
LEVEL III	Build It								■	
LEVEL III	Balloon Bubbles (ii)									
LEVEL IV	Talking Toy (iii)			■					■	
LEVEL IV	Thirty Seconds			■				■		
LEVEL IV	Balloon Bubbles (iii)			■						
LEVEL IV	Story Line			■				■		
LEVEL IV	Post Box							■		

The shaded boxes indicate the other skill areas addressed by this activity.

BODY LANGUAGE

Level	Activity	Auditory Memory	Phonological Awareness	Speaking	Understanding	Listening and Attention	Turn-Taking	Feelings and Emotions	Confidence and Independence	Awareness of Others
LEVEL I	Watch Me!						■			
LEVEL I	Thumbs Up!				■	■				
LEVEL I	Magic Messages						■			
LEVEL I	Magic Mime				■					■
LEVEL I	Magic Movements					■			■	
LEVEL II	Magic Box (i)						■	■		
LEVEL II	One Thing for Another						■		■	
LEVEL II	Elves and Goblins								■	
LEVEL II	Follow My Leader								■	
LEVEL II	Grandmother's Footsteps									
LEVEL III	Magic Movements (ii)								■	■
LEVEL III	Gesture Sentences				■			■	■	■
LEVEL III	Statues								■	■
LEVEL III	Special Sitting				■					■
LEVEL III	Watch Me! (ii)					■				
LEVEL IV	Fidget Fiends					■				
LEVEL IV	Mime Story				■	■				■
LEVEL IV	Magic Movements (iii)					■			■	
LEVEL IV	Magic Box (ii)								■	
LEVEL IV	Hurrah-Boo!				■		■		■	

The shaded boxes indicate the other skill areas addressed by this activity.

AWARENESS OF OTHERS

The shaded boxes indicate the other skill areas addressed by this activity.

	Activity	Auditory Memory	Phonological Awareness	Speaking	Understanding	Listening and Attention	Turn-Taking	Feelings and Emotions	Confidence and Independence	Body Language
LEVEL I	Empty Chair						■		■	
	Who Has Gone?						■		■	
	About Us			■	■					
	Who is Asleep?									■
	Listen and Jump!				■		■			
LEVEL II	Lining Up				■		■			
	Ask Me!			■	■		■			
	Find Me!				■	■	■			
	Yummie Yuckie			■			■	■		
	Squashed Bananas			■		■	■			■
LEVEL III	Guess Who?				■		■		■	
	News			■			■		■	
	Alien Visitor (i)			■			■			
	Four Corners of the Earth			■		■	■			
	New Kid						■	■		
LEVEL IV	Alien Visitor (ii)			■		■	■			
	Help!			■	■		■			■
	Gift Box			■			■			■
	Wash the Puppy			■			■			■
	Who Am I?			■		■	■			

CONFIDENCE AND INDEPENDENCE

		Auditory Memory	Phonological Awareness	Speaking	Understanding	Listening and Attention	Turn-Taking	Feelings and Emotions	Body Language	Awareness of Others
LEVEL I	Roundabout					■				
	Take Rabbit				■		■			
	Fetch It!			■	■	■				
	Choosing Chairs			■		■				
	What You Need	■								
LEVEL II	Pass the Beanbag						■			■
	Hot and Cold					■				■
	Sergeant Major			■		■				
	1-2-3 Choose!	■			■					
	Auntie Jean's Birthday			■	■					■
LEVEL III	Lions and Tigers			■					■	
	Scavenge Hunt				■	■				
	Collections	■			■					
	Specially Me			■						■
	Messengers	■			■					
LEVEL IV	Job to Job									
	Egg Timer									
	Relay Race						■			■
	Tower of Babel			■	■					
	Put the Tail on the Donkey									■

The shaded boxes indicate the other skill areas addressed by this activity.

FEELINGS AND EMOTIONS

The shaded boxes indicate the other skill areas addressed by this activity.

Level	Activity	Auditory Memory	Phonological Awareness	Speaking	Understanding	Listening and Attention	Turn-Taking	Body Language	Confidence and Independence	Awareness of Others
LEVEL I	Happy–Sad						▓	▓		
	If You're Happy…						▓	▓		
	How do I Feel?			▓			▓	▓		▓
	Thank You–No Thank You			▓			▓	▓		▓
	Zig's Day			▓				▓		▓
LEVEL II	Angry–Scared						▓	▓		▓
	Good People			▓				▓		▓
	Who Let the Cat Out?			▓	▓	▓				
	All Change!			▓	▓			▓		
	Blues for the Blues							▓	▓	▓
LEVEL III	In a Dark Dark Cave			▓				▓	▓	
	Neighbours			▓			▓			▓
	Compliments Bag			▓			▓		▓	▓
	Sandcastle Game			▓			▓		▓	▓
	Pampering Pets			▓	▓		▓			▓
LEVEL IV	Worry Beads			▓	▓				▓	▓
	Prize Draw			▓	▓				▓	
	Party Plan			▓	▓		▓		▓	▓
	Listen to My Voice					▓		▓		▓
	Spin-a-Word			▓		▓				

RESOURCES LANGUAGE AND LITERACY

UNDERSTANDING

Level	Activity	Auditory Memory	Phonological Awareness	Speaking	Listening and Attention	Body Language	Turn-Taking	Feelings and Emotions	Confidence and Independence	Awareness of Others
LEVEL I	If!				■					
LEVEL I	Category Bingo				■					
LEVEL I	Listen and Colour	■			■					
LEVEL I	Find Zig						■		■	
LEVEL I	What Can it Be?				■		■			
LEVEL II	Red and Yellow Counters	■			■					
LEVEL II	Zig's Tea	■			■					
LEVEL II	Farmer Fred				■					
LEVEL II	Parrot Hunt	■			■		■			
LEVEL II	Where's Granny Going?				■					
LEVEL III	Musical Messages				■		■			
LEVEL III	Three Clues			■	■					
LEVEL III	Art Attack	■			■					
LEVEL III	Rats' Tails (i)				■					■
LEVEL III	Once Upon A Time	■			■					
LEVEL IV	It's a Funny World				■				■	
LEVEL IV	Guess Who?			■	■					■
LEVEL IV	Listen and Draw	■			■					
LEVEL IV	Rats' Tails (ii)				■					■
LEVEL IV	Work it Out!	■			■					

The shaded boxes indicate the other skill areas addressed by this activity.

DEVELOPING BASELINE COMMUNICATION SKILLS

LISTENING AND ATTENTION

Level	Activity	Auditory Memory	Phonological Awareness	Speaking	Understanding	Body Language	Turn-Taking	Feelings and Emotions	Confidence and Independence	Awareness of Others
LEVEL I	Listening Walk								■	■
LEVEL I	Hunt the Sound		■							
LEVEL I	Go Game				■					
LEVEL I	Guess the Instrument	■								
LEVEL I	Musical Bumps								■	■
LEVEL II	Where am I?									
LEVEL II	Copy Cat	■								
LEVEL II	Listening Feet				■		■			
LEVEL II	Mystery Sounds									
LEVEL II	Mousie-Mousie®									
LEVEL III	High or Low?				■					■
LEVEL III	Threes	■								
LEVEL III	Fruit Salad	■								
LEVEL III	Oranges and Lemons	■			■				■	
LEVEL III	Count the Bears									
LEVEL IV	Zoo Game				■					
LEVEL IV	Finders Keepers	■								
LEVEL IV	You Got it Wrong!				■					
LEVEL IV	Colouring Rainbows							■		
LEVEL IV	Bandstand									■

The shaded boxes indicate the other skill areas addressed by this activity.

SPEAKING

Level	Activity	Auditory Memory	Phonological Awareness	Listening and Attention	Understanding	Body Language	Turn-Taking	Feelings and Emotions	Confidence and Independence	Awareness of Others
LEVEL I	Something's Missing			■					■	
LEVEL I	Big Green Apples			■	■					
LEVEL I	Circus Act						■		■	
LEVEL I	Do As I Say						■		■	
LEVEL I	Raindrops			■						
LEVEL II	Zig's Day			■						
LEVEL II	Colour Families			■			■			
LEVEL II	George the Giant			■			■			
LEVEL II	Zig's Picnic								■	
LEVEL II	Picture Partners				■				■	
LEVEL III	Obstacle Course	■		■	■					
LEVEL III	Our Own Story			■					■	
LEVEL III	Spot the Difference						■			
LEVEL III	Disguises						■		■	■
LEVEL III	Only One Left			■	■		■			
LEVEL IV	Zig at the Zoo			■					■	
LEVEL IV	Make It Up								■	
LEVEL IV	Gold Crowns			■	■				■	
LEVEL IV	Oops!			■	■					
LEVEL IV	Imagine It								■	

The shaded boxes indicate the other skill areas addressed by this activity.

AUDITORY MEMORY

Level	Activity	Listening and Attention	Phonological Awareness	Speaking	Understanding	Body Language	Turn-Taking	Feelings and Emotions	Confidence and Independence	Awareness of Others
LEVEL I	Shopping Game (i)	■			■					
LEVEL I	Farmers and Mechanics	■			■					
LEVEL I	Magic Passwords			■						
LEVEL I	Things to Do	■			■					
LEVEL I	Postbag	■								
LEVEL II	Shopping game (ii)	■			■					
LEVEL II	Magic Pictures	■			■					
LEVEL II	Chinese Whispers (i)	■								■
LEVEL II	Colour Jumping	■			■					
LEVEL II	Magic Carpets	■			■				■	
LEVEL III	Pets' Corner	■		■						
LEVEL III	Whose News?	■						■		
LEVEL III	Chinese Whispers (ii)	■							■	■
LEVEL III	Ring Me Up!	■							■	
LEVEL III	Postbag	■								
LEVEL IV	Parrots	■							■	
LEVEL IV	Pantomimes	■				■				
LEVEL IV	Fact Finder	■								
LEVEL IV	Matching Pictures (ii)	■			■					
LEVEL IV	Jack and Jill	■		■						

The shaded boxes indicate the other skill areas addressed by this activity.

PHONOLOGICAL AWARENESS

		Auditory Memory	Listening and Attention	Speaking	Understanding	Body Language	Turn-Taking	Feelings and Emotions	Confidence and Independence	Awareness of Others
LEVEL I	Clap Your Name		■				■			■
	Switch Me Off			■						
	Pass on the Code		■				■			
	Poems Please!	■	■						■	
	Hurry Them Up!		■							
LEVEL II	Come When I Clap		■						■	
	My Mistake									
	Beanbags									■
	Incy-Wincy-Mincy-Pincy			■						
	Pitter Patter									■
LEVEL III	On the Beat		■				■			
	Cats and Rats									
	Sound Ladders			■						
	Rhyming Families			■					■	
	Stepping Stones									
LEVEL IV	Prove It									■
	Rhyme the Number			■						
	TV Tongue Twisters			■					■	
	Odd One Out			■					■	
	Bob's Bunkbed									

The shaded boxes indicate the other skill areas addressed by this activity.

Developing Baseline Communication Skills
PUPIL RECORD SHEETS

Personal and Social

294 / Personal and Social Development
295 / Language and Literacy

PERSONAL AND SOCIAL DEVELOPMENT

Child's Name	Skill Area																				
	Turn-Taking				Body Language				Awareness of Others				Confidence and Independence				Feelings and Emotions			
	I	II	III	IV	I	II	III	IV	I	II	III	IV	I	II	III	IV	I	II	III	IV

LANGUAGE AND LITERACY

Child's Name	Skill Area															
	Understanding				Listening and Attention				Speaking				Auditory Memory				Phonological Awareness			
	I	II	III	IV	I	II	III	IV	I	II	III	IV	I	II	III	IV	I	II	III	IV

© C Delamain & J Spring 2000 Photocopiable